Science in Uniform, Uniforms in Science

Historical Studies of American Military and Scientific Interactions

Edited by
Margaret Vining
Barton C. Hacker

Martin Gordon, Consulting Editor

SMITHSONIAN NATIONAL MUSEUM OF AMERICAN HISTORY, WASHINGTON, D.C.
in association with
THE SCARECROW PRESS, INC.
Lanham, Maryland • Toronto • Plymouth, UK
2007

SCARECROW PRESS, INC.

Published in the United States of America
by Scarecrow Press, Inc.
A wholly owned subsidiary of
The Rowman & Littlefield Publishing Group, Inc.
4501 Forbes Boulevard, Suite 200, Lanham, Maryland 20706
www.scarecrowpress.com

Estover Road
Plymouth PL6 7PY
United Kingdom

British Library Cataloguing in Publication Information Available

Library of Congress Cataloging-in-Publication Data

Science in uniform, uniforms in science : historical studies of American military and scientific interactions / edited by Margaret Vining, Barton C. Hacker.
p. cm.
Includes bibliographical references and index.
ISBN-13: 978-0-8108-5991-3 (pbk. : alk. paper)
ISBN-10: 0-8108-5991-2 (pbk. : alk. paper)
1. Military research—United States—History. 2. Science and state—United States—History. I. Vining, Margaret. II. Hacker, Barton C., 1935–

U393.5S35 2007
355'.070973—dc22 2007008128

∞™ The paper used in this publication meets the minimum requirements of American National Standard for Information Sciences—Permanence of Paper for Printed Library Materials, ANSI/NISO Z39.48-1992.
Manufactured in the United States of America.

Contents

Foreword

Margaret Vining
Barton C. Hacker

This collection of essays owes its existence to the fortuitous conjunction of two events. The first was a temporary exhibition at the Smithsonian's National Museum of American History in Washington, DC, that opened in October 2002. Sixteen months later, in January 2004, the annual meeting of the American Historical Association came to the same city. The AHA seemed to offer a perfect venue for the exhibit's final public program, a symposium on the historic links between America's armed forces and the development of American science and technology.

The exhibition was entitled West Point in the Making of America, 1802–1918. Sponsored by the U.S. Army, it commemorated the bicentennial of the U.S. Military Academy at West Point. Rather than recount the academy's history, however, this exhibit focused on the lives and work of a select group of West Point graduates, some famous, others less well known, in the context of American national development from the beginning of the nineteenth century through World War I. One of the exhibit's central themes was the significant part West Pointers played in the creation of American science and engineering. An extraordinary display of objects, such as natural history specimens sent by antebellum soldier-explorers in the West to the newly formed Smithsonian Institution, augmented the biographical narratives with visual and material historical evidence.

West Point in the Making of America was developed during a transitional period of exhibit production at the National Museum of American History. Traditional methods of exhibit development relied almost entirely on the efforts of individual curators, who did most of the fund-raising and work themselves, with an assist from technical experts where required. The new concept invoked a team model in which curators joined managers, designers, specialists, and technicians. The West Point exhibit bowed to the past in the sense that it was a narrowly focused temporary show—fifteen months—for which curator-historians supplied story line and intellectual content. But the team approach relieved curators of many of their traditional responsibilities. A project manager orchestrated the work of a team of professionals, coordinating the exhibit process from curatorial concept through design, artifact conservation, production, installation, and public programs. In the team approach, a dedicated staff and budget sustained each activity. Included in the budget was money for public programs: lectures, presentations, tours, and other activities intended to expand public awareness of some of the exhibit's broader implications.

When the American Historical Association announced that the theme of its 2004 annual meeting would be "War and Peace: History and the Dynamics of Human Conflict and Cooperation," it seemed the obvious venue for the exhibit's final public program. We planned a symposium around the theme of historical interactions between America's armed forces and American science and technology. Our call for papers drew an unexpectedly large number of fine abstracts, allowing us to assemble an ambitious four-session, sixteen-paper proposal, which the AHA accepted for its program under the title of "The Armed Forces Interacting with American Science and Technology, from the Revolution to the Twenty-first Century." The symposium was held at the museum over the middle two days of the four-day meeting.

The special Smithsonian-AHA symposium offered an unusual opportunity to present new research by both established and younger scholars on the interaction of the armed forces and American science. At the National Museum of American History, this theme resonated against the backdrop of the West Point exhibition where army officers broke new ground for military-civil collaboration in the formative decades of American science and technology. Not all those who participated in the symposium were able to prepare articles for this volume. It nonetheless represents an impressive cross-section of work being done on an important but too-often-overlooked aspect of American history.

1

Science, Technology, and American Military Institutions, from the Revolution through the Cold War

Barton C. Hacker

The Industrial Revolution and the American Revolution were coeval. Despite breaking with England, the former colonies continued to share a common language and culture with their former ruler. Continuing immigration from the British Isles augmented the new republic's access to the birthplace of industrialism and related sources of military technological innovation. This was already an old American pattern. Throughout the colonial period the European heritage loomed large, but transplanted institutions rarely flourished unchanged. American environments consistently, if not uniformly, altered knowledge and practice carried from the Old World, in military matters as in others. American engineering, scientific, and military institutions drew on European sources, but always to some lesser or greater degree adjusted them to New World needs.[1]

Gun making along traditional lines flourished in the colonies, and colonial naval stores became a major resource of British sea power. Interactions of military and naval engineering with other preindustrial American institutions, even as limited as they were, had virtually no counterpart in science. Colonial science foreshadowed the observational preoccupations that became so marked a feature of nineteenth-century American science. It also remained distinctly provincial, if not backward, and had no discernible effect on military institutions.[2] Much the same was true of medicine, still in the eighteenth century much more art and craft than science. Military surgeons and nurses might accompany larger expeditions, but hardly affected operations. Like all early modern armies, both British and American forces in the Revolutionary War relied on camp followers as nurses. Medical services, too, remained little changed from earlier practices. Although the Continental army acquired a medical department, the long-standing pattern of ten deaths from disease for every death in battle persisted.[3] Yet the promise of something better emerged from the successful smallpox inoculation program in Washington's army described here by Ann Becker.

Engineering in the United States formally began with the founding of the United States Military Academy at West Point in 1802. The new school focused on training engineers, but military technical training proved to have many uses, outside the army as well as within. West Point graduates helped found civil engineering in the United States, both as builders and as teachers. Military engineers also filled key managerial roles in both civil and military enterprises. Railroad corporations owed them as great a debt as the government arsenals that pioneered interchangeable parts manufacturing. By the mid-nineteenth century the traffic between Europe and America was ceasing to be one way.[4]

During the early nineteenth century, American science remained more strongly oriented toward observation than experimentation. Exploring and mapping westward across the continent during the first half of the nineteenth century, army-led expeditions supported that bent. They provide a stream of specimens and notes, drawings and photographs, maps and descriptions that promoted vigorous American contributions to such primarily observational fields as astronomy, natural history, the earth sciences, and anthropology. Mid-century also saw the U.S. Navy assume a major role in exploration.[5] In this volume, K. B. Sterling examines two army and two navy exploring expeditions in greater detail.

The American Civil War straddled two ages. It was both the last great preindustrial war and the first major war of the industrial age.[6] On the eve of war the United States was not yet an industrial nation, but the process of industrialization was well under way. A key factor was steam power, which was used to produce and transport goods. Military engineers had once again assumed a prominent role. West Point provided many of the railroad builders, and Annapolis later began to bolster the ranks of steam engineers. Industrial capacity attained new levels of military significance as transportation improved, but in this, as in many other respects, the Civil War was distinctly transitional, and that strongly affected its conduct.[7]

During the late nineteenth century, United States armed forces became chiefly observers of major changes in military technology and organization taking place in Europe. Both army and navy reverted to prewar levels and tasks within a few years of the Civil War's close. Policing the frontier and patrolling distant stations again became their main functions.[8] Military-scientific cooperation in exploration revived during the 1870s and 1880s, though to a much lesser extent than before the war. The Weather Service, the Geological Survey, the Coast Survey, even the Naval Observatory, all of which had for varying lengths of time worked under military auspices, became purely civilian enterprises. The Corps of Engineers remained an army operation, but grew increasingly civilianized.[9] In this volume, James Fleming shows how the late-nineteenth-century Weather Service was shaped by army concerns—and how subsequent military concerns continued to shape the development of meteorology, not always for the better.

In contrast to the earlier nineteenth century, the later saw industrialism playing a key role in nurturing applied science, with major consequences for the conduct of military affairs. New science-based industries—chemical, electrical, metallurgical—began to affect weaponry more directly than ever before and to exert increasingly significant battlefield effects. Old patterns hardly vanished, nor did their impact even much lessen. Outsiders continued to invent and promote new or improved weapons, including the decisive military technological innovations of the late nineteenth century—magazine rifles, machine guns, and quick-firing field artillery. And yet there was also something new. Industrial research laboratories had begun to search for innovations scientifically. Smokeless powder arose from such research, pointing clearly toward future lines of development.[10]

The new relation between industry and technology transformed navies. Naval contracts, once largely limited to times of imminent or actual warfare, became peacetime staples in the late nineteenth century. Undergoing rapid technological change that overwhelmed government shipyards, the new steam-and-steel navies required increasingly closer cooperation with industry to remain competitive. All the major powers embarked on ambitious shipbuilding programs. During the 1880s and 1890s, the U.S. Navy began rebuilding itself as a modern sea force. Naval revival fostered economic growth, particularly in steel and related industries, and drew heavily on the growing ability of scientifically oriented engineering to produce specified armor and equipment to order.[11] It also contributed to the rise of navalism based on an exciting new theory of sea power. America's imperial venture opened auspiciously with the successful naval war against Spain. The naval arms race acquired its most potent symbol at the dawn of the new century with the British launching of HMS *Dreadnought* in 1906.[12] Walter Pittman in his article describes the political aspects of this naval technological revolution in America.

World War I became the first great industrial war, imposing extraordinary demands on the economies and societies of the belligerents. Manufacturing and logistics mattered more than all other aspects of war making. Every state had to reorganize its people and resources to sustain its home front, cope with shortages, maintain production, and keep supplies flowing to the war fronts. The new weapons demanded ammunition in staggering quantities, but that was only the first crisis. Armies so huge required supplies of every kind on a formerly unimagined scale. "War economy" and "home front" entered the lexicon for the conversion of industrial capacity, for the reorientation of civic life, for the concentration of all resources toward fighting total war.[13] Bureaucratic management formerly limited to individual business firms or government agencies now applied to the entire state and displaced free-market capitalism at the center of war economies. Science took its place among the activities the state sought to control, with varying success.[14]

Meanwhile, though, less direct consequences of scientific mobilization for World War I may have been more significant. Military techniques of classifying jobs and sorting personnel required little adjustment to suit the precepts of scientific management.[15] Colleges became enthusiastic centers for military training and education. During the era of World War I, American universities used military models in organizing research laboratories and adapted their curricula to meet military demands. They also borrowed military test methods to evaluate their students.[16] Mark Soderstrom here offers a close examination of the University of Minnesota's role in designing IQ tests for the army, then using those same tests to shape its student body after the war.

Medical services, too, displayed strikingly greater efficiency in 1917–1918 than they had in earlier American wars. Reorganization after the Spanish-American War contributed to the improvement, as did the opportunity to prepare vouchsafed by America's long-anticipated decision to enter the war. Nothing resembling the epidemics that raced through 1898 training camps afflicted the recruits of World War I. The record remained equally good overseas, at least until the fall of 1918, when a virulent and highly contagious strain of influenza devastated the warring armies and civilian populations as well. Even so, disease accounted for but half of troop deaths in France, despite appalling conditions on the Western Front.[17] American medical achievements during World War II were even more remarkable. Disease, which before the twentieth century had always accounted for the vast majority of deaths in war, was all but eliminated as a significant factor in military mortality.[18] John Morra follows the career of Albert Sabin to show the role of epidemiological research in reducing the impact of disease during World War II and how Sabin's wartime experience contributed to his more famous postwar achievement in the conquest of polio.

From 1940 onward, the United States devised new forms for mobilizing science that succeeded beyond almost anyone's expectations. Relationships among science, engineering, and military institutions emerged from World War II utterly transformed. The reluctant military innovators of the later nineteenth century became enthusiastic seekers of technological novelty in the later twentieth. Perhaps more significant, they began finding money to support the scientific research and development engineering, as well as merely the testing and procurement, new weapons required. Since the seventeenth century, scientists had justified their requests for state support by claiming the ability to enhance war-making potential. Generally, such claims proved will-o'-the-wisps, seeming near but rarely quite within reach.

In the United States during World War II, science seemed at last to have achieved its long-professed goal.[19]

Perhaps even more important, American industry proved able to produce the novelties generated by science in decisive quantities. The mobilization of industry, like the mobilization of science, was notably more successful in World War II than in World War I. The aircraft industry, in particular, pioneered new methods of organizing production that played a decisive role, not only in World War II itself, but also in the transformation of industry that followed the war.[20] Robert Ferguson explores the wartime organizational innovations in aircraft manufacturing and suggests some of their long-term significance.

During World War II, science in the broad sense that has become normal in this century—including engineering, technology, and medicine as well as basic and applied research—effectively met military needs as never before. Whether the military co-opted science or scientists bent the military to their own purposes, though much discussed, is probably irresolvable. Both sides paid and both benefited.[21] Equally important, military officers and government officials reversed their long-standing and often justified skepticism about scientists' ability to serve military needs efficiently. World War II seemed to prove beyond any reasonable doubt that liberal funding could convert scientific theory into battlefield asset fast enough to change the course of war in progress. It also demonstrated that military needs and desires could control and direct scientific research, in effect confirming the prophecies of prewar advocates of mechanization and air power.[22]

Nuclear weapons and the still unsolved problems of air warfare brought science directly into military strategic planning. The civilian scientific advisers of the war years came to look a good deal more like staff officers without uniforms after the war. Beginning with the air force–sponsored Rand Corporation, the practice spread throughout the military establishment. Physicists, operational analysts, and other civilian experts applied quantitative methods to strategic problems and assumed a growing share of the province formerly occupied exclusively by general staffs. At the same time, well-funded military interests increasingly shaped the scientific agenda.[23]

The unexpectedly rapid Soviet achievement of a nuclear test explosion in 1949 inspired the American decision to proceed with a crash program to develop thermonuclear weapons (hydrogen or H-bombs). It also led to the formation of another major national weapons laboratory (Livermore) and, in conjunction with war in Korea, a major increase in federal funding for weapons research and development. The network of government-owned, contractor-operated national laboratories became major centers of research in the physical sciences and, later, in the biomedical sciences as well.[24]

The equally unexpected Soviet success in launching the first artificial earth satellite, *Sputnik,* in 1957, likewise had major repercussions for military-scientific research. The most direct result was another quasi-military federal science agency, the National Aeronautics and Space Administration (NASA). Established in 1958, it took over the National Advisory Committee for Aeronautics (NACA), as well as several military rocket projects. Like the Atomic Energy Commission (AEC), it relied heavily on academic and industrial contractors for much of its R&D. Other less direct consequences of *Sputnik* were the institutionalization of science advice to the president, Pentagon reorganization to strengthen defense science and advanced research, and greatly expanded funding for education related to national defense.[25]

Sputnik spurred U.S. missile development and the elaboration of strategic nuclear deterrence. The creation of what became known as the strategic triad was already well under way, reaching fruition between 1955 and 1960 with the deployment of three major new weapon systems able to deliver nuclear weapons over intercontinental distances. Long-range jet-propelled manned bombers became the first leg of the triad as B-52s began reaching operational air force units in 1955. Intercontinental ballistic missiles (ICBMs) required major breakthroughs in guidance systems and nuclear warhead design; the first generation of liquid-propellant ICBMs, Atlas and Titan, were deployed in the late 1950s, with the second-generation solid-propellant Minuteman following in the early 1960s. The final leg of the triad became operational in 1960 with the first patrol of a nuclear-powered submarine armed with nuclear-tipped Polaris missiles.[26] In his essay for this volume, Barton Hacker explores the transformation of submarines and their development as undersea missile platforms.

Although nuclear weapons and their delivery systems remained, along with the specter of a third world war, the focus of strategic thinking and public concern, they had virtually no impact on the actual conduct of war in the last half of the twentieth century. Grandiose schemes for the large-scale automation of defense decision making from SAGE (semiautomatic ground environment) air defense system in the 1960s to the Strategic Defense Initiative (Star Wars) antiballistic missile system in the 1980s have produced meager returns for vast costs.[27]

More successful have been efforts to enhance the capabilities of individual weapons and weapon systems, though not without raising serious concerns about skyrocketing costs and operational reliability. Armed forces now deploy weapons of vastly greater power and accuracy, communications far more extensive and reliable, and enormously more efficient systems of command and control. What has mainly mattered in the field is the astonishing increase in the potency of conventional weaponry and support systems. "Conventional" with respect to modern weaponry is simply taken to mean non-NBC (nuclear, biological, chemical).[28]

Underlying these developments is the explosive growth of microelectronics fueled to a significant degree by military money. The history of electronic computing offers particularly cogent instances of the new relationship between science and the military. Integrated into antiaircraft artillery systems that included radar guidance and sensor activation, electromechanical computers had achieved a measure of combat success during World War II. They had also proved of considerable help in calculating firing tables for artillery and completing other tedious computational chores. The promise of fully electronic computing fired military imaginations and loosened military purse strings after the war.[29]

Science-based technological innovation promised greatly enhanced command and control of military forces. Information warfare joined precision-guided weapons as a major focus of thinking about the structure of armed forces in the late twentieth century. Satellite early warning, surveillance, and reconnaissance systems were major factors in maintaining nuclear stability during the Cold War. Among the most significant innovations in military technology since World War II were communication satellites, both military and civilian. From small beginnings in the 1960s, they have come to dominate long-distance global message traffic.[30]

Reliable and nearly instantaneous global telecommunications have transformed the military chain of command, a mixed blessing for commanders in the field, as naval commanders learned with the introduction of radio at the beginning of the twentieth century.[31] And just as radio vastly increased the coordination of forces in World War II, the newer forms of telecommunication brought field commands still more firmly under central direction. Just how firmly, and sometimes uncomfortably, emerges clearly in Gen. Wesley Clark's account of his role as NATO commander in the Balkan Wars of the late 1990s.[32]

Both the United States and the Soviet Union deployed satellite systems for a variety of other useful military functions. Several of these systems, or their civilian counterparts, can provide valuable information for nonmilitary purposes as well, though usually at lower resolutions. Navigational satellites have been scarcely less important than communication satellites in their effects on military operations. Accurate global positioning systems allow commanders to see the exact location of their units at any moment, lifting much of the fog of war.[33] The obvious importance of weather and accurate ranging to military operations underscores the significance of meteorological and geodetic systems; like communication systems, they have found burgeoning civil as well as military uses.[34] Erik Conway shows how military concerns shaped the development of satellite meteorological research.

Engineering was the first, and until modern times the only, military science. Engineers built fortifications and devised such ingenious machines as catapults to attack them. During the nineteenth century, engineering diversified, specialized, and professionalized. Engineering education shifted from shop floor to classroom, engineering practice from cut-and-try to analysis and computation. Some kinds of engineering acquired the patina of science, as well as a measure of control over innovation and production. Achievements such as these in time overcome the resistance of military planners threatened by too-rapid technological change. In the twentieth century, and especially since 1950 in the United States, military demands backed by huge defense budgets restored much of engineering to the military fold, in practice if not in name: contracts and specifications have displaced uniforms and commands as the effective methods of exerting control.

Science was less tightly tied to military institutions than was engineering, though the dream (or nightmare) of science-based breakthroughs in weaponry has a long history. Relationships have, in any event, grown steadily closer as science demonstrated its military utility, beginning with eighteenth-century advances in ballistics. Utility grew more pronounced during the nineteenth century. Although most technological innovation still depended chiefly on trial and error, applied science and systematic innovation began to achieve impressive results. Just as engineering was becoming more systematic and mathematical, science was tending toward greater practicality. Differences between applied science and scientific engineering became ever less distinct.

By the end of the twentieth century, science spread its mantle not only across military technology and much engineering, but over virtually every area of military enterprise. Activities as diverse as personnel management and strategic planning have joined technology as apparently subject to the rule of science. Engineering has, at the same time, expanded its domain to include the manipulation of people and social groups as well as materials and processes. Military usage now consistently lumps scientific research with engineering development and product testing in a single enterprise. At the beginning of the twenty-first century, science as well as engineering has taken on a distinct military cast.

NOTES

1. Neil Longley York, *Mechanical Metamorphosis: Technological Change in Revolutionary America* (Westport, CT, and London: Greenwood Press, 1985); George D. Moller, *American Military Shoulder Arms*, vol. 1, *Colonial and Revolutionary War Arms* (Niwot: University Press of Colorado, 1993).

2. Brooke Hindle, *The Pursuit of Science in Revolutionary America, 1735–1789* (Chapel Hill: University of North Carolina Press, 1956); Silvio A. Bedini, *Thinkers and Tinkers: Early American Men of Science* (New York: Charles Scribner's Sons, 1975); Nathan Reingold, *Science, American Style* (New Brunswick, NJ: Rutgers University Press, 1991).

3. Mary C. Gillett; *The Army Medical Department, 1775–1818* (Washington, DC: Center of Military History, 1981); Mary T. Sarnecky, *A History of the U.S. Army Nurse Corps* (Philadelphia: University of Pennsylvania Press, 1999), 1–8; Holly A. Mayer, *Belonging to the Army: Camp Followers and Community during the American Revolution* (Columbia: University of South Carolina Press, 1996), 219–23.

4. James L. Morrison Jr., *The Best School in the World: West Point, the Pre–Civil War Years, 1833–1866* (Kent, OH: Kent State University Press, 1986); Todd Shallat, *Structures in the Stream: Water, Science, and the Rise of the U.S. Army Corps of Engineers* (Austin: University of Texas Press, 1994); Charles Francis O'Connell, "The Corps of Engineers and the Rise of Modern Management, 1827–1856," in *Military Enterprise and Technological Change: Perspectives on the American Experience*, ed. Merritt Roe Smith (Cambridge, MA: MIT Press, 1984), 87–116; Merritt Roe Smith, "Military Entrepreneurship," in *Yankee Enterprise: The Rise of the American System of Manufacturing*, ed. Otto Mayr and Robert C. Post (Washington, DC: Smithsonian Institution Press, 1981), 63–102.

5. William H. Goetzmann, *Army Exploration in the American West, 1803–1863* (New Haven, CT: Yale University Press, 1959); Frank N. Schubert, *Vanguard of Expansion: Army Engineers in the Trans-Mississippi West, 1819–1879* (Washington, DC: Office of the Chief of Engineers, 1980); Vincent Ponko Jr., *Ships, Seas, and Scientists: U.S. Naval Exploration and Discovery in the Nineteenth Century* (Annapolis, MD: Naval Institute Press, 1974).

6. Stig Förster and Jörg Nagler, eds., *On the Road to Total War: The American Civil War and the German Wars of Unification, 1861–1871* (Washington, DC: German Historical Institute; Cambridge: Cambridge University Press, 1997); Williamson A. Murray, "The Industrialization of War, 1815–1871," in *The Cambridge Illustrated History of Warfare: The Triumph of the West*, ed. Geoffrey Parker (Cambridge: Cambridge University Press, 1995), 216–41.

7. Paul A. C. Koistinen, *Beating Plowshares into Swords: The Political Economy of American Warfare, 1606–1865* (Lawrence: University Press of Kansas, 1996), part 2, "Economic Mobilization for War—The Transitional Stage"; Louis C. Hunter, *A History of Industrial Power in the United States, 1780–1930*, vol. 2, *Steam Power* (Charlottesville: University Press of Virginia, 1985); Robert V. Bruce, *The Launching of Modern American Science, 1846–1876* (New York: Alfred A. Knopf, 1987).

8. Walter Millis, *Arms and Men: A Study of American Military History* (New York: G. P. Putnam's Sons, 1956), chap. 3, "The Managerial Revolution"; Jerry K. Sweeney et al., *A Handbook of American Military History: From the Revolutionary War to the Present* (Boulder, CO: Westview Press, 1996), chap. 3, "The Reform Period: 1866–1914"; James M. Morris, *America's Armed Forces: A History* (Englewood Cliffs, NJ: Prentice Hall, 1991), chap. 6, "Military Arm of an Expanding Nation, 1865–1914."

9. Barton C. Hacker, "Engineering a New Order: Military Institutions, Technical Education, and the Rise of the Industrial State," *Technology and Culture* 33 (1993): 1–27; Peter Karsten, "Militarization and Rationalization in the United States, 1870–1914," in *The Militarization of the Western World*, ed. John R. Gillis (New Brunswick, NJ: Rutgers University Press, 1989), 30–44.

10. Frederick Myatt, *The Illustrated Encyclopedia of 19th Century Firearms: An Illustrated History of the Development of the World's Military Firearms during the 19th Century* (New York: Crescent Books, 1979); Geof Bowker, "Manufacturing Truth: The

Development of Industrial Research," in *A History of Scientific Thought: Elements of a History of Science*, ed. Michel Serres, trans. from the French (Cambridge, MA: Blackwell, 1995), 583–610; Sandra Lee Norman, "Guncotton to Smokeless Powder: The Development of Nitrocellulose as a Military Explosive" (PhD diss., Brown University, 1988).

11. Benjamin Franklin Cooling, *Gray Steel and Blue Water Navy: The Formative Years of America's Military-Industrial Complex, 1881–1917* (Hamden, CT: Archon Books, 1979); Thomas J. Misa, *A Nation of Steel: The Making of Modern America, 1865–1925* (Baltimore, MD: Johns Hopkins University Press, 1995); John D. Alden, "Growth of the New American Navy (1888–1898)," in *Naval Engineering and American Seapower*, ed. Randolph W. King, Prescott Palmer, and Bruce I. Meader (Baltimore, MD: Nautical & Aviation Publishing Company of America, 1989), 33–64.

12. Robert Seager II, "Alfred Thayer Mahan: Christian Expansionist, Navalist, and Historian," in *Admirals of the New Steel Navy: Makers of the American Naval Tradition, 1880–1930*, ed. James C. Bradford (Annapolis, MD: Naval Institute Press, 1990), 73–96; Jon Sumida, *In Defence of Naval Supremacy: Finance, Technology, and British Naval Policy, 1889–1914* (London: Unwin Hyman, 1989).

13. John F. Godfrey, *Capitalism at War: Industrial Policy and Bureaucracy in France, 1914–1918* (Leamington Spa, UK: Berg, 1987); Roger Chickering and Stig Förster, eds., *Great War, Total War: Combat and Mobilization on the Western Front, 1914–1918* (Washington, DC: German Historical Institute; Cambridge: Cambridge University Press, 2000); Paul A. C. Koistinen, *Mobilizing for Modern War: The Political Economy of American Warfare, 1865–1919* (Lawrence: University Press of Kansas, 1997), part 2, "Economic Mobilization for World War I."

14. A. Hunter Dupree, *Science in the Federal Government: A History of Policies and Activities to 1940* (Cambridge, MA: Harvard University Press, 1957) chap. 16, "The Impact of World War I, 1914–1918"; William H. McNeill, *The Pursuit of Power: Technology, Armed Force, and Society since A.D. 1000* (Chicago: University of Chicago Press, 1983), chap. 9, "World Wars of the Twentieth Century."

15. David F. Noble, *America by Design: Science, Technology, and the Rise of Corporate Capitalism* (New York: Alfred A. Knopf, 1977); Penn Borden, *Civilian Indoctrination of the Military: World War I and Future Implications for the Military-Industrial Complex* (New York: Greenwood Press, 1989).

16. Paul Davis Chapman, "Schools as Sorters: Testing and Tracking in California, 1910–1925," *Journal of Social History* 14 (1981): 701–17; Joel H. Spring, "Psychologists and the War: The Meaning of Intelligence in the Alpha and Beta Tests," *History of Education Quarterly* 12 (1972): 3–15; Richard T. von Mayrhauser, "The Manager, the Medic, and the Mediator: The Clash of Professional Psychological Styles and the Wartime Origins of Group Mental Testing," in *Psychological Testing and American Society, 1890–1930*, ed. Michael M. Sokal (New Brunswick, NJ: Rutgers University Press, 1987), 128–57.

17. Mary C. Gillett, *The Army Medical Department, 1865–1917* (Washington, DC: Center of Military History, 1995); John M. Barry, *The Great Influenza: The Epic Story of the Deadliest Plague in History* (New York: Viking Penguin, 2004); Carol R. Byerly, *Fever of War: The Influenza Epidemic in the U.S. Army during World War I* (New York and London: New York University Press, 2005).

18. Peter Neushul, "Fighting Research: Army Participation in the Clinical Testing and Mass Production of Penicillin During the Second World War," in *War, Medicine*

and Modernity, ed. Roger Cooter, Mark Harrison, and Steve Sturdy (Stroud, UK: Sutton, 1998), 203–24; Albert E. Cowdrey, *Fighting for Life: American Military Medicine in World War II* (New York: Free Press, 1994).

19. A. Hunter Dupree, "The *Great Instauration* of 1940: The Organization of Scientific Research for War," in *The Twentieth-Century Sciences: Studies in the Biography of Ideas*, ed. Gerald Holton (New York: W. W. Norton, 1972), 443–67; Carroll Pursell, "Science Agencies in World War II: The OSRD and Its Challengers," in *The Sciences in the American Context: New Perspectives*, ed. Nathan Reingold (Washington, DC: Smithsonian Institution Press, 1979), 359–78.

20. Paul A. C. Koistinen, *Arsenal of World War II: The Political Economy of American Warfare, 1940–1945* (Lawrence: University Press of Kansas, 2004); I. B. Holley Jr., *Buying Aircraft: Materiel Procurement for the Army Air Forces*, United States Army in World War II: Special Studies (Washington, DC: Center of Military History, 1964).

21. U.S. Cong., House Comm. on Science and Technology, Task Force on Science Policy, *A History of Science Policy in the United States, 1940–1985*, by Jeffrey K. Stine, Science Policy Study Background Report No. 1, 99th Cong., 2nd Sess., September 1986; Everett Mendelsohn, Merritt Roe Smith, and Peter Weingart, eds., *Science, Technology and the Military*, 2 vols. (Dordrecht: Kluwer Academic, 1988); Gregg Herken, *Cardinal Choices: Presidential Science Advising from the Atomic Bomb to SDI* (New York: Oxford University Press, 1992).

22. Paul Forman and José M. Sánchez-Ron, eds., *National Military Establishments and the Advancement of Science and Technology* (Dordrecht: Kluwer Academic, 1996); Paul Forman, "Behind Quantum Electronics: National Security as Basis for Physical Research in the United States, 1940–1960," *Historical Studies in the Physical and Social Sciences* 18 (1987): 149–229; Chandra Mukerji, *A Fragile Power: Scientists and the State* (Princeton, NJ: Princeton University Press, 1989); John Tirman, ed., *The Militarization of High Technology* (Cambridge, MA: Ballinger, 1984).

23. Fred Kaplan, *The Wizards of Armageddon* (New York: Simon & Schuster, 1983); Gregg Herken, *Counsels of War* (New York: Alfred A. Knopf, 1985); Paul Dickson, *Think Tanks* (New York: Atheneum, 1971).

24. Dan Kevles, "K_1S_2: Korea, Science, and the State," in *Big Science: The Growth of Large-Scale Research*, ed. Peter Galison and Bruce Hevly (Stanford, CA: Stanford University Press, 1992), 312–33; David Alan Rosenberg, "The Origins of Overkill: Nuclear Weapons and American Strategy, 1945–1960," *International Security* 7 (1983): 3–71; Herbert York, *The Advisors: Oppenheimer, Teller, and the Superbomb* (1976; reprint, Stanford, CA: Stanford University Press, 1989); Peter Galison and Barton Bernstein, "In Any Light: Scientists and the Decision to Build the Superbomb, 1952–1954," *Historical Studies in the Physical and Biological Sciences* 19 (1989): 267–347.

25. Alan J. Levine, *The Missile and Space Race* (Westport, CT: Praeger, 1994); Robert A. Divine, *The Sputnik Challenge* (New York: Oxford University Press, 1993); Walter A. McDougall, *The Heavens and the Earth: A Political History of the Space Age* (New York: Basic Books, 1985).

26. Edmund Beard, *Developing the ICBM: A Study in Bureaucratic Politics* (New York: Columbia University Press, 1976); Jacob Neufeld, *The Development of Ballistic Missiles in the United States Air Force, 1945–1960* (Washington, DC: Office of Air Force History, 1989); Harvey M. Sapolsky, *The Polaris System Development: Bureaucratic and Programmatic Success in Government* (Cambridge, MA: MIT Press, 1972).

27. Paul N. Edwards, *The Closed World: Computers and Politics in Cold War America* (Cambridge, MA: MIT Press, 1996); Donald R. Baucom, *The Origins of SDI, 1944–1983* (Lawrence: University Press of Kansas, 1992); Sanford Lakoff and Herbert F. York, *A Shield in Space? Technology, Politics, and the Strategic Defense Initiative* (Berkeley and Los Angeles: University of California Press, 1989); B. Bruce-Briggs, *The Shield of Faith: A Chronicle of Strategic Defense from Zeppelins to Star Wars* (New York: Simon & Schuster, 1988).

28. Guy Hartcup, *The Silent Revolution: The Development of Conventional Weapons, 1945–85* (London: Brassey's, 1993); Bengt Anderberg and Myron L. Wolbarsht, *Laser Weapons: The Dawn of a New Military Age* (New York: Plenum, 1992); Mary Kaldor, *The Baroque Arsenal* (New York: Hill & Wang, 1981); Geoffrey Kemp, Robert L. Pfaltzgraff Jr., and Uri Ra'anan, eds., *The Other Arms Race: New Technologies and Non-Nuclear Conflict* (Lexington, MA: Lexington Books, 1975); Julian Perry Robinson, "Qualitative Trends in Conventional Munitions: The Vietnam War and After," in *The World Military Order: The Impact of Military Technology on the Third World*, ed. Mary Kaldor and Asbjørn Eide (New York: Praeger, 1979), 64–109.

29. David A. Mindell, *Between Human and Machine: Feedback, Control, and Computing before Cybernetics* (Baltimore, MD: Johns Hopkins University Press, 2002); I. Bernard Cohen, "The Computer: A Case Study of Support by Government, Especially the Military, of a New Science and Technology," in *Science, Technology and the Military*, ed. Everett Mendelsohn, Merritt Roe Smith, and Peter Weingart, 2 vols. (Dordrecht: Kluwer Academic, 1988), 1:119–54.

30. Edward A. Comor, *Communication, Commerce, and Power: The Political Economy of America and the Direct Broadcast Satellite* (New York: St. Martin's Press, 1998); Andrew J. Butrica, *Beyond the Ionosphere: Fifty Years of Satellite Communication* (Washington, DC: National Aeronautics and Space Administration, 1997).

31. Susan J. Douglas, "The Navy Adopts the Radio, 1899–1919," in *Military Enterprise and Technological Change: Perspectives on the American Experience*, ed. Merritt Roe Smith (Cambridge, MA: MIT Press, 1985), 117–74. See also Tony Devereux, *Messenger Gods of Battle: Radio, Radar, Sonar: The Story of Electronics in War* (London: Brassey's, 1991); Arthur Hezlet, *Electronics and Sea Power* (New York: Stein & Day, 1975).

32. Wesley K. Clark, *Waging Modern War: Bosnia, Kosovo, and the Future of Combat* (New York: Public Affairs, 2001). Cf. Michael O'Hanlon, *Technological Change and the Future of Warfare* (Washington, DC: Brookings Institution Press, 2000), chap. 3, "Sensors, Computers, and Communications"; Chris Hables Gray, *Postmodern War: The New Politics of Conflict* (New York: Guilford Press, 1997); Gene I. Rochlin, *Trapped in the Net: The Unanticipated Consequences of Computers* (Princeton, NJ: Princeton University Press, 1997), chap. 11, "C^3I in Cyberspace"; C. Kenneth Allard, *Command, Control, and the Common Defense* (New Haven, CT: Yale University Press, 1990).

33. Michael Russell Rip and James M. Hasik, *The Precision Revolution: GPS and the Future of Aerial Warfare* (Annapolis, MD: Naval Institute Press, 2002); Pradipta Shome, *History and Chronology of GPS* (Washington, DC: U.S. Office of Commercial Space Transportation, 2002); George W. Bradley III, "Origins of the Global Positioning System," in *Technology and the Air Force: A Retrospective Assessment*, ed. Jacob Neufeld, George M. Watson Jr., and David Chenoweth (Washington, DC: Air Force

History & Museums Program, 1997), 245–54; Bradford W. Parkinson, "A History of Satellite Navigation," *Navigation* 42 (1995): 109–64; Scott Pace et al., *The Global Positioning System: Assessing National Policies* (Santa Monica, CA: Rand Critical Technologies Institute, 1995), app. B, "GPS History, Chronology, and Budgets."

34. Deborah Jean Warner, "Political Geodesy: The Army, the Air Force, and the World Geodetic System of 1960," *Annals of Science* 59 (2002): 363–89; R. Cargill Hall, *A History of the Military Polar Orbiting Meteorological Satellite Program* (Washington, DC: National Reconnaissance Office, 2001); P. Krishna Rau, *Evolution of the Weather Satellite Program in the U.S. Department of Commerce: A Brief Outline* (Washington, DC: National Oceanic and Atmospheric Administration, 2001).

2

Inoculation in Washington's Army: The Battle against Smallpox

Ann M. Becker

"The Smallpox! The smallpox! What shall we do with it?" So mourned John Adams as he contemplated the collapse of the Northern Army's Canadian campaign in June 1776.[1] This lamentation illustrates the scope and magnitude of the contemporary response to the problem of smallpox during the American Revolution. The disease had an enormous negative effect on the Canadian campaign and complicated strategic planning during the Siege of Boston. American recruitment was stymied, desertion rates increased, and commanding officers had to proceed with inadequate forces due to the prevalence of smallpox among the troops. Lack of immunity among the soldiery severely compromised the ability of the American military to function during the early years of the American Revolution.

The need to control the crippling presence of smallpox among Revolutionary soldiers led directly to the development of a comprehensive military policy of inoculation and increased the popularity of this procedure with the American citizenry. Although this tool to prevent smallpox had been available for decades, the American colonies lacked the public health resources to take advantage of it. Military discipline enabled George Washington to impose the use of large-scale mandatory inoculation in 1777, which removed the threat of smallpox from the army and led to its acceptance and wider use by the general population.

SMALLPOX IN AMERICA

Smallpox was universally feared in North America—during the seventeenth and eighteenth centuries, an epidemic in one or more sections of the country was expected periodically.[2] British historian Thomas Babington Macaulay asserted "that disease . . . was then the most terrible of all the ministers of death . . . always present . . . tormenting with constant fears all whom it had not yet stricken, leaving on those whose lives it spared the hideous traces of its power."[3] Early medical historian James G. Mumford noted that "at the beginning of that era, of all the foes our ancestors faced—hardship, famine, pestilence, Indian and foreign wars—the most dreaded was small-pox."[4] Mortality from naturally acquired smallpox ranged from 15 to 50 percent.

In Europe smallpox was primarily an endemic disease, where the infection was always present within a certain population, particularly in urban areas, and generally suffered in childhood. In rural America, however, smallpox appeared periodically in epidemic form, where outbreaks of the disease occurred as a vulnerable population was born or moved into an area. The appearance of smallpox within their community struck fear into the hearts of colonists, as most Americans were liable to catch the disease and understood its dangerous implications.[5]

Extremely infectious, the disease spread directly from human sources. Although the mechanism of contagion had not yet been proven scientifically, colonials understood that exposure to smallpox led to infection. Physical contact allowed the *variola* virus (which caused the disease) to enter the body through a skin abrasion. Airborne particles from any contaminated person or surface could be ingested through respiration or by mouth. Once exposed, a person not protected by an existing immunity would almost certainly develop smallpox. Medical historian John Duffy noted that close to 100 percent of those exposed to smallpox without prior inoculation or immunization acquired the infection.[6]

The distinctive characteristics of the disease included headache, chills, backache, high fever, vomiting, and anxiety, as well as an easily recognizable rash and distinctive pock. The incubation period was about twelve days from initial exposure until the eruption of symptoms and the onset of the contagious phase of the illness. Residual effects of the disease were horrible. Smallpox could pit the skin grotesquely, cause eyebrows and lashes to fall out, leave blindness in its wake, and result in scarring so severe as to close up the nostrils. The disfigurement left by the disease led to social ostracism and even suicide for some victims.[7]

Long-standing colonial policies of enforced quarantine and isolation were not able to prevent epidemics from recurring in the American colonies on a regular basis. Outbreaks of smallpox were recurrent, devastating, and frequent, with attendant high fatality rates.[8] Inoculation was the first reliable preventive measure available to combat the disease. New England minister Cotton Mather became aware of its benefits when he read a medical paper published in the *Philosophical Transactions of the Royal Society* in 1714. The paper substantiated the claims of Mather's Negro slave Onesimus, who had described undergoing the procedure in Africa. A widely practiced folk custom in

China, India, other Asiatic countries, the Middle East, and Africa since the first century, inoculation was introduced in England in 1720 and Boston in 1721.[9]

The medical procedure of inoculation consisted of physically placing contagious matter from an active lesion into a bodily incision. A mild, though contagious, case of the disease generally resulted, followed by lifelong immunity.[10] Fatality rates from inoculation ranged from 1 to 9 percent, depending on the health and age of the patient and the method used. Naturally acquired smallpox was much more severe and had significantly higher mortality rates.[11]

The decision to infect oneself with a potentially fatal, often disfiguring, and highly contagious disease was not made lightly. Many colonists rejected voluntarily introducing smallpox into a community, since improper management of inoculation could result in the virulent spread of the disease.[12] Most rural New England towns and villages were able to avoid infection altogether for long periods of time due to their relative isolation, so the deliberate introduction of smallpox into a community through use of inoculation was not popular.[13] Once the disease appeared among a vulnerable population, whether naturally or by inoculation, smallpox usually spread dangerously and quickly.

Those inoculated often did not, or could not, maintain isolation or restrict their activity while contagious, and frequently infected those they came in contact with.[14] Due to this danger, restrictive laws governed inoculation and local governments commonly permitted its use only after isolation and other precautions had failed and the epidemic spread of smallpox was inevitable.[15] Local officials, constrained by public opinion and controlled by popular vote, had little impetus to institute the wider use of inoculation.[16] During the decades prior to the Revolution, bans on the procedure were lifted only temporarily in times of crisis.

According to medical historian John Blake, the "individual's desire to guard himself or his family against a deadly disease conflicted with the community's same need for protection." Unless carefully managed and strictly controlled, inoculation threatened many with the disease while offering protective immunity for only a few fortunate individuals. Poorer residents were unable to afford either the fees (which ranged from two to five pounds and included room and board, preparatory medications and nursing care) or the recuperative time (about three weeks) spent in isolation. Preventive medical care, then, was generally available only to wealthier Americans, although municipalities often made efforts to inoculate the poor during smallpox epidemics.[17]

SMALLPOX IN TIME OF WAR

Wartime conditions created a new crisis in controlling this highly infectious virus. As he assumed command of the American forces in July 1775,

Washington was keenly aware that the largest threat to the health of the fledgling army massed outside Boston was the smallpox epidemic within the city. The disease flourished when large groups of previously unexposed populations converged, as they did in army camps during the Revolutionary War.[18] Dr. Benjamin Rush later noted that epidemics arose with the "sudden assembling of a great number of persons together of different habits and manners, such as the soldiers of the American army were in the years of 1776 and 1777."[19] Newly recruited soldiers, recently arrived from outlying rural areas, were well fed and in good health, but liable to contract the *variola* virus.[20] Although port cities frequently suffered epidemics due to recurrent exposures of contagion from incoming ships, rural communities were relatively safe from smallpox for long periods of time. This lack of previous exposure to smallpox as children, however, rendered many soldiers extremely likely to develop this disease once exposed, and this complicated military efforts.[21]

Shortly after arriving at Cambridge, Washington informed the Continental Congress that he had been "particularly attentive to the least Symptoms of the Small Pox," and asserted, "We shall continue the utmost Vigilance against this most dangerous Enemy."[22] With a relatively stable troop population massed in stationary camps and medical professionals readily accessible, military commanders isolated smallpox cases and strictly enforced isolation and notification requirements. Washington restricted camp access, maintained a quarantine hospital, and checked refugees to avoid an epidemic.[23] A smallpox survivor himself, the general understood the effect the disease might have on the health of his soldiers and the ability of his army to function.[24] The general's efforts outside Boston successfully prevented an epidemic among his troops during the siege, but smallpox remained a threat to the American army.

Control of the disease became increasingly difficult once the British evacuated Boston in March 1776. Boston had been plagued with outbreaks of the disease since the British arrived there in 1774, and efforts at containment had proved futile.[25] The day the British left, Boston selectmen alerted General Washington that "small-pox was in many places in the Town, and it would not be prudent for Persons who had not had the Distemper to venture in."[26] Washington sent only immune troops to take possession of the Heights outside Boston in a calculated attempt to control the spread of this contagious disease. The general also continued to rigorously enforce isolation of its victims.[27] These measures, however, could not contain the explosion of smallpox cases in the town that spring and summer.

Unable to forestall the epidemic, civil and military authorities ultimately resorted to mass inoculation to fight smallpox in Boston. With a severe epidemic underway, the Massachusetts legislature temporarily lifted its ban on the procedure for twelve days in July 1776, during which time nearly five thousand were inoculated. Military physician Dr. James Thacher received orders "to inoculate for the small-pox, all the soldiers and inhabitants in town, as a general infection of this terrible disease is apprehended."[28] In the eyes of Moses Morse, a soldier in the ranks, "Boston is become a hospital with the smallpox." Abigail Adams took advantage of the opportunity and underwent inoculation with her children. "I now date from Boston," she wrote John, "where I yesterday arrived and was with all 4 of our Little ones inoculated for the small pox." Guards were posted at roads and ferries throughout the summer to contain the infection within the town. Efforts to halt the spread of disease were ultimately successful in Boston, largely due to the massive and effective use of inoculation.[29]

Smallpox prevention continued to be a priority when Washington moved his base of operations to New York in March of 1776. A quarantine hospital, guarded and staffed by soldiers with immunity, had already been established there, and inoculations were being performed. The danger of uncontrolled contagion was ever present, however, and fearing the spread of smallpox in the city and among his troops, Washington ended use of the preventive procedure upon his arrival in Manhattan. If careful controls and restrictions were not enforced, inoculation tended to disburse the *variola* virus. Anxious to avoid this, the general promised, "[A]ny disobedience to this order will be most severely punished." Military and civilian authorities continued to isolate smallpox victims, but a private physician in New York was jailed for inoculating soldiers.[30] As in Boston, efforts to contain smallpox here were successful, although in this case isolation rather than mass inoculation was the tool used to contain smallpox.

CRISIS AT QUEBEC

The military situation in Canada was far different, and the impact of smallpox much more serious. As the Northern Army laid siege to Quebec during the winter and spring of 1776, the disease erupted among the troops. The first cases appeared in early December.[31] Without adequate medical care or housing and suffering from logistical and supply problems, military leaders could not control the spread of smallpox. Few of the troops serving in Canada had been previously exposed to the deadly contagion, and were likely to contract it once exposed. The *variola* virus flourished in the chaotic conditions before Quebec, and American soldiers serving there knew that if they did not carry immunity, they were almost certain to develop the infection. Methods of containment and prevention previously used were futile. Isolation was impossible, and conditions were not conducive to the safe and effective use of inoculation. Unwilling to risk being infected with smallpox naturally, soldiers began to inoculate themselves contrary to military orders.

In January, Massachusetts private James Melvin noted, "more taken with smallpox, and we expect it will be a general disorder, for we are very thick, nasty and lousy." Containment of the infection proved impossible, given the difficult and crowded living conditions. By February fifty men were hospitalized with smallpox, and within two weeks an additional fifty succumbed to the disease. Inexperienced army medical personnel failed to recognize the seriousness of the situation until it was too late to stop the spread of smallpox, and logistical problems made the implementation of a structured inoculation program impossible.[32] The disease made rapid advances among the troops, yet inoculation was prohibited because of its tendency to spread smallpox if adequate controls were not put in place.[33]

Gen. Benedict Arnold issued orders against the procedure on 11 February, but soldiers continued to inoculate themselves.[34] Fearing that "spreading the S-Pox . . . will probably prove the entire ruin of the army," Arnold issued additional orders on 15 March, imploring his officers to "take all possible care to prevent [the spread of smallpox]." He forbade the "Surgeons of the Army . . . under the severest penalty, to inoculate any person." Punishment included the cashiering of officers and court-martial for the rank and file.[35] How effective these edicts could have been is debatable, given the overwhelming fear the disease held for the men and the fact that inoculation would almost certainly result in a mild case of smallpox and the sought-after immunity. Threatening to dismiss soldiers and officers from a dangerous and deadly situation simply did not have the desired effect—the mortal danger of smallpox was so real to the soldiers, they refused to desist.

Official inoculation efforts were not begun until the disease had made significant inroads among the soldiers in Canada. The constant influx of reinforcements, most lacking immunity to *variola*, compounded the problem. A majority of the new recruits were not immune to the disease, and sickened quickly.[36] Gen. David Wooster commented that the soldiers sent to Canada "took the smallpox as they arrived, either by inoculation, or the natural way."[37] Charged with investigating events in Canada after the war, Thomas Jefferson discovered that "the first recruits after the defeat came in about 6 weeks. . . . As fast as they came they were laid up with the smallpox. Might be about 27 or 2800 [total]. . . . Of these might be not more [than] 800 or 900 effective."[38] Nearly 60 percent of the reinforcements contracted smallpox via inoculation. Whole regiments submitted to the procedure, rendering them useless for military duty. "From the 1st of January to the 1st of March, we have never had more than seven hundred effective men on the ground, and frequently not more than five hundred," wrote General Arnold to Silas Deane.[39] Instead of adding to American military capabilities in Canada, reinforcements only served to add an additional burden to an already struggling army—caring for men in various stages of inoculated or naturally contracted smallpox.

The haphazard and rampant use of illicit inoculation created a spiraling sequence of events—as soldiers submitted to the operation to protect themselves, the contagion spread. With heightened fear and an increased chance of natural infection, the use of unsupervised self-inoculation exploded, complicating an already difficult strategic situation. American military authorities attempted to protect the Northern Army by prohibiting inoculation, yet the soldiers were so anxious to avoid taking smallpox the more severe "natural" way, they insisted on inoculating themselves without using proper medical techniques or precautions. Soldiers under inoculation could not be effectively quarantined and were contagious, physically debilitated, and unable to perform their duties for up to three weeks. By April, nearly one-half of the force before Quebec was ill with the disease. Quarantine was impossible to enforce.[40] Reinforcements continued to pour into Canada, and the contagion spread unchecked. Statistical data indicates that by May, American forces in Quebec numbered 8,048, with most susceptible to or infected with smallpox.[41]

Widespread and uncontrolled inoculation and the constant influx of vulnerable recruits newly exposed to the contagion fueled the spread of smallpox among the troops. With his ongoing efforts to prevent unsupervised inoculation fruitless, and hoping to curtail the effect of the disease, on 15 May Arnold recommended inoculation of troops before they were dispatched to Canada, and on 16 May the general, with the acquiescence of the Congressional Commission, finally instituted a short-lived policy permitting the procedure.[42] The effort was too little, too late.

In May 1776 Gen. John Thomas, a medical doctor and accomplished military commander, arrived in Quebec to assume command of the American forces.[43] He found thousands ill with smallpox, and had no time to safely inoculate his men. Unable to control the procedure and anxious to maintain an effective and healthy fighting force for his planned campaign, Thomas countermanded Arnold's last order, which had allowed inoculation.[44] In an attempt to control the rampant spread of the disease, and facing an imminent British attack, Thomas issued new orders (which carried the death penalty) forbidding the operation. As an example to his troops, though he was not immune, the general refused to inoculate himself.[45]

Fear of the disease and the likelihood of death if infected naturally, however, rendered injunctions against inoculation useless. "The smallpox is an infinite detriment to the service; notwithstanding which and the most express orders to the contrary, both officers and soldiers privately inoculate themselves," lamented Thomas.[46] Rapidly changing events and smallpox had conspired to leave the American forces in Canada too weak to sustain the onslaught of British regulars, forcing an ignominious and deadly retreat. General Thomas was stricken with smallpox in May and died during the American retreat, only one month after his arrival in Quebec.[47]

Anecdotal evidence places the blame for the devastating defeat in Canada squarely on smallpox. Northern Army commander Gen. Horatio Gates conceded that his army had been "entirely ruined with smallpox."[48] John Adams lamented, "Our misfortunes in Canada are enough to melt a heart of stone. The small-pox is ten times more terrible than Britons, Canadians, and Indians together."[49] Gen. John Sullivan, who replaced Thomas after his death, complained to his commanding officer, Gen. Philip Schuyler, "that infernal disorder, the small-pox, has ruined our army."[50] Ultimately, the congressional committee ordered to investigate the military loss in Canada concluded "that a still greater, and more fatal, source of misfortune has been, the prevalence of the small pox in that army."[51]

The danger to Washington's army posed by the disease continued unabated through 1776, and an effective policy to eliminate smallpox became crucial for American military success. Washington hoped his Northern Army commanders would have the same success he had in Boston in containing the disease. On 7 July the general optimistically assured Governor Jonathan Trumbull, "If proper precautions are taken, the small-pox may be prevented from spreading" among the troops returning from Canada. Washington explained that "this done at Cambridge, and I trust will be contrived by General Schuyler and Gates, who are well apprized of the fatal Consequences that may attend Its infecting the whole Army."[52] Ultimately Washington's efforts bore fruit. The Northern Army was clear of the disease by September 1776, when General Gates declared, "Thank Heaven, the small-pox is totally eradicated from amongst us, not I can assure you, without much vigilance and authority being previously exerted."[53]

INOCULATION ADOPTED

The successful—though limited—use of troop inoculations in the face of epidemic smallpox after the Boston and Canadian campaigns in 1776 led directly to Washington's decision to utilize this preventive procedure more widely the following year. The earlier inoculations, accomplished with the help of governmental authorities as smallpox threatened to spread out of control in both civilian and military populations, had been successful. General Washington soon realized that simply reacting to the presence of smallpox would undermine his efforts to maintain and develop an effective military organization. A preemptive strike against the deadly disease was vital. Inoculation offered the only effective solution to what had become a crisis within the American army.

After the ignominious end of the Canadian campaign, visions of an army crippled by smallpox haunted the American commander in chief. Though he had used selective inoculation, isolation, and immune troops effectively earlier in the war, these measures were no longer sufficient. A new strategy to protect his troops and sustain the Revolution emerged—systematic mass troop inoculation. Linking his decision directly to events in Canada, Washington asserted, "The deplorable and melancholy situation, to which one of our Armies was reduced last Campaign by the small Pox . . . has determined me . . . to introduce inoculation immediately." In January 1777, with the approval of the Continental Congress, Washington resolved to begin mass inoculations in his army.[54]

The commander in chief took a great military risk by instituting this policy. Use of the procedure, though valuable in combating smallpox among the soldiery, carried with it significant drawbacks. While the general advocated the adoption of "vigorous methods" to "remove the Infected and Infection before we feel too sensible the effects," he realized that the use of inoculation, isolation, and quarantine could be "disagreeable and inconvenient to Individuals."[55] Mortality from inoculation was low, yet soldiers did die as a result of the procedure. In addition, while soldiers recuperated from induced smallpox, the army would be severely weakened for a lengthy period of time, and the British would have had a significant strategic advantage if aware of the debilitated condition of their enemy.[56] Washington cautioned Dr. William Shippen in January to keep "the matter as secret as possible," and instituted inoculation procedures immediately.[57]

The general, however, soon had second thoughts about this aggressive and innovative course of action. Perhaps fearful of the consequences should the British discover his plan, and given the extensive movement of troops that winter—particularly the influx of "Eastern" recruits, who had "not yet had the Small pox"—Washington decided to "check, not spread the Infection" a short time later. Natural cases of smallpox would likely increase if those inoculated were not effectively quarantined until all possibility of contagion had subsided.

The army was still at great risk, however, if the smallpox infection was allowed to proliferate. It soon became clear that inoculation was no longer optional, in spite of the procedure's dangerous tendency to spread the *variola* virus. The disease could not be controlled unless military authorities actively pursued innovative medical tactics. By February 1777 the general found new resolve, and "determined . . . to inoculate all the troops" at Morristown as well as "the Recruits as fast as they come into Philadelphia."[58]

Washington, with the aid of civil authorities, now acted quickly to implement this new military medical policy designed to eliminate the threat of smallpox. Doctors were dispatched to Trenton and Morristown in New Jersey; to Fishkill, Ticonderoga, and the Hudson Highlands in New York; to Bethlehem, Newtown, and Philadelphia in Pennsylvania; and to several locations in Connecticut, where mass inoculation of soldiers and recruits were carried out.[59] On 20 February the Continental Congress directed the

Maryland State Assembly to provide such medications as required "to inoculate the Continental troops in this town." In April the Congress resolved that "Major General Schuyler . . . hasten the march of Carolina Continental troops" and halt these soldiers "at Dumfries, Colchester, and Alexandria in Virginia, there to pass through inoculation."[60] Divisions submitted to the operation en masse at five- to six-day intervals. The army used guarded private homes and churches as treatment and isolation centers.[61]

Soldiers, camp followers, and proximate citizens all needed protection from smallpox. Washington tried to assure his men "that their Families will be under not the smallest danger of catching the small pox. I have taken every possible care of them . . . to prevent the Infection's spreading."[62] Americans were increasingly exposed to the disease due to wartime demographic shifts and troop movements as hostilities continued. Civilians were often inoculated along with the army, although not always willingly. Local historians credited Rev. Timothy Johnes with "persuading his parishioners of the efficacy of this operation," and aiding Washington's efforts to "stay the ravages of the terrible disease." Despite the military precautions, some evidence indicates that the contagion escaped Continental army camps, as smallpox infected residents of Princeton and Morristown in the spring 1777.[63]

The army's use of this medical treatment proved highly successful and the mortality rate for these supervised inoculations averaged less than 1 percent. This compared very favorably with the usual 15 to 50 percent mortality from naturally occurring smallpox. Inoculated smallpox was considerably milder, and had a shorter recuperative period as well. Clearly, the benefits to the army greatly outweighed the risks. By late 1777 the Continental army's policy of inoculation was well established, using safe techniques and careful controls.[64] Englishman Daniel Sutton had popularized a technique using shallow incisions in the 1760s, which lessened the introduction of secondary bacterial infection seen in earlier methods utilizing deeper subcutaneous cuts. Known as the Dimsdale method in the American colonies, this safer way of inoculating eliminated preparatory purging and bleeding, and used contagious matter taken from an inoculated donor early in the course of illness. These revised practices lowered the mortality rate considerably.[65]

INOCULATION VALIDATED

The use of inoculation to prevent the spread of smallpox was a significant and innovative advance in military medical care.[66] Washington removed a major threat to the health of his soldiers and protected the overall strength and reliability of the Continental army by systematically inoculating his troops.[67] Once citizens and soldiers recovered from the milder inoculated smallpox and acquired immunity, they were permanently "freed of the apprehensions of that fatal distemper."[68] Military leaders were able to manage their strategic planning without fearing the untimely explosion of a smallpox epidemic. From a public health perspective, compulsory army inoculations also helped encourage the civilian population to use this preventive procedure as well. Though the technology and knowledge to prevent smallpox was available, the American colonies lacked the public health resources to take advantage of this valuable medical procedure. Resistance to inoculation faded as it became impossible to avoid smallpox contagion and military use of the procedure proved its effectiveness. Inoculation gained popularity with American citizens with the greatly increased likelihood of acquiring the infection during the war and as they became increasingly familiar with the protection it afforded.[69]

Despite occasional setbacks and difficulties instituting his inoculation policy, Washington's efforts to eliminate smallpox from his army were successful. His position as commander in chief, and the support of civilian authorities, allowed him to institute the widespread use of a dangerous, though effective, measure to control a deadly disease. In the face of military disaster, public health concerns took precedence over civil liberties, and local authorities permitted and encouraged the use of inoculation to protect American soldiers and citizens. The army's use of mass inoculation proved the procedure's usefulness and effectiveness in preventing this dangerous disease, leading to its increased popularity and availability during the war and in the decades after the Revolution. Under the pressure of war, military leaders were able to solve a medical problem that civil authorities had struggled with for decades.

The positive effects of Washington's use of inoculation were recognized even before the end of the war. In 1781 Rush noted, "It must afford no small pleasure to a benevolent mind . . . to reflect, that the small-pox which once proved equally fatal to thousands, has been checked in its career, and in a great degree subdued by the practice of inoculation."[70] By the 1780s this operation was accepted as the standard protection against smallpox, replaced only with the introduction of Dr. Edward Jenner's cowpox vaccine in 1798, which offered protection against smallpox without contagion or illness.[71] Washington's successful effort to control smallpox saved his army and helped protect all Americans against the deadly disease.

NOTES

1. Peter Force, *American Archives Consisting of a Collection of Authentick Records, State Papers, Debates and Letters and Other Notices of Public Affairs, the Whole Forming*

a Documentary History of the Origin and Progress of the North American Colonies (Washington, DC: M. St. Clair & Peter Force, 1837–53), 4:6, 1083, Extract of a letter from John Adams, 26 June 1776.

2. For a description of the physical effects of smallpox, including incubation period and symptoms, see Elizabeth A. Fenn, "Biological Warfare in 18th Century North America: Beyond Jeffrey Amherst," *Journal of American History* 86 (2000): 1522. See also Patricia A. Watson, *The Angelic Conjunction: Preacher-Physicians of Colonial New England* (Knoxville: University of Tennessee Press, 1991), 41; Jonathan B. Tucker, *Scourge: The Once and Future Threat of Smallpox* (New York: Atlantic Monthly Press, 2001), 9. For a description of public reaction to smallpox, see Daniel J. Boorstin, *The Americans: The Colonial Experience* (New York: Random House, 1958), 214–20.

3. Thomas Babington Macaulay, *History of England* (London: Macmillan, 1914), as quoted in Solon S. Bernstein, "Smallpox and Variolation: Their Historical Significance in the American Colonies," *Journal of the Mount Sinai Hospital* 18 (1951): 228.

4. James G. Mumford, *A Narrative of Medicine in America* (Philadelphia and London, 1903), as quoted in John T. Barrett, "The Inoculation Controversy in Puritan New England," *Bulletin of the History of Medicine* 12 (1942): 170.

5. Watson, *Angelic Conjunction*, 15. Watson wrote: "In England, smallpox had been endemic, a common yet dreaded affliction to which most of the population was exposed as children. In New England, the disease was sporadic, and the isolation of the individual towns often prevented it from being transported from one area to another. With each successive generation, the population grew steadily, rapidly producing new groups of non-immune children. The effect was a regular increase in the frequency of epidemics." For a detailed discussion of smallpox in Britain, see S. R. Duncan, Susan Scott, and C. J. Duncan, "Smallpox Epidemics in Cities in Britain," *Journal of Interdisciplinary History* 25 (1994): 255–71.

6. Watson, *Angelic Conjunction*, 41; Tucker, *Scourge*, 9; Nancy Tomes, *The Gospel of Germs: Men, Women and the Microbe in American Life* (Cambridge, MA, and London: Harvard University Press, 1998), 3; John Duffy, *Epidemics in Colonial America* (Baton Rouge: Louisiana State University Press, 1953), 24; J. Worth Estes, "'A Disagreeable and Dangerous Employment': Medical Letters from the Siege of Boston, 1775," *Journal of the History of Medicine* 31 (1976): 275.

7. Richard B. Stark, "Immunization Saves Washington's Army," *Surgery, Gynecology and Obstetrics* 144 (1977): 425; Hugh Thursfield, "Smallpox in the American War of Independence," *Annals of Medical History* 2 (1940): 313; Francis D. Moore, "Zabdiel Boylston and Colonial Surgery," *Bulletin of the American College of Surgeons* 68 (May 1983): 13. For mortality estimates, see C. W. Dixon, *Smallpox* (Boston: Little, Brown, 1962), 171; Ola Elizabeth Winslow, *A Destroying Angel: The Conquest of Smallpox in Colonial Boston* (Boston: Houghton Mifflin, 1974), 61.

8. Watson, *Angelic Conjunction*, 41; Tucker, *Scourge*, 9; Moore, "Zabdiel Boylston and Colonial Surgery," 13; Estes, "Siege of Boston," 275; Dixon, *Smallpox*, 171; Duffy, *Epidemics*, 24.

9. Henry R. Viets, "Some Features of the History of Medicine in Massachusetts During the Colonial Period," *Isis* 23 (1935): 399–400; John B. Blake, *Public Health in the Town of Boston, 1630–1822* (Cambridge, MA: Harvard University Press, 1959), 54; Winslow, *Destroying Angel*, 35. Cotton Mather survived smallpox at the age of fif-

teen, and his family experienced its severe effects during the 1702 Boston epidemic. The word "quarantine" is derived from the Italian for "forty," indicating the amount of time victims of contagious disease were to be restricted.

10. J. Worth Estes, "The Practice of Medicine in 18th Century Massachusetts: A Bicentennial Perspective," *New England Journal of Medicine* 305, no. 18 (October 1981): 1043; Stanhope Bayne-Jones, *The Evolution of Preventive Medicine in the United States Army, 1607–1939* (Washington, DC: Office of the Surgeon General, Department of the Army, 1968), 15; Donald R. Hopkins, "Benjamin Waterhouse, the 'Jenner' of America,'" *American Journal of Tropical Medical Hygiene* 5, pt. 2 supp. (September 1977): 1060; Elizabeth A. Fenn, *Pox Americana: The Great Smallpox Epidemic of 1775–1782* (New York: Hill & Wang, 2001), 32–33. Inoculation deliberately introduced the infection to the patient, which resulted in a mild case of smallpox resulting in lifelong immunity. This phenomenon has not yet been explained scientifically. The patient was contagious while undergoing the procedure. Vaccination, or immunization, results in a very mild, nonsymptomatic infection, which is not contagious, but only bestows temporary immunity. Death is rare from vaccination and infrequent with inoculation.

11. Dixon, *Smallpox*, 243–45; F. J. Spencer, "Medical Affairs at the Battle of Yorktown," *Bicentennial Programs* (Evansville: St. Mary Hospital, Indiana Graduate Medical Center, 1976), 59.

12. Duffy, *Epidemics*, 38; E. H. Kass, "A Brief Perspective on the Early History of American Infectious Disease Epidemiology," *Yale Journal of Medical Biological Medicine* 60 (July–August 1987): 342; James H. Cassedy, *Medicine in America: A Short History* (Baltimore, MD, and London: Johns Hopkins University Press, 1991), 13.

13. Blake, *Public Health*, 112.

14. Fenn, *Pox Americana*, 40–41. Abigail Adams described attending meeting every day while under inoculation in Boston in 1776.

15. John B. Blake, "The Medical Profession and Public Health in Colonial Boston," *Bulletin of the History of Medicine* 26 (1952): 225–26; J. Worth Estes, "Medical Skills in Colonial New England," *New England Historical and Genealogical Register* 134 (1980): 268; Cassedy, *Medicine in America*, 13; Blake, *Public Health*, 108–109; Abbas M. Behbehani, *The Smallpox Story in Words and Pictures* (Kansas City: University of Kansas Medical Center, 1988), 33; James E. Gibson, "The Role of Disease in the 70,000 Casualties of the American Revolution," *Transactions and Studies of the College of Physicians of Philadelphia* 17 (December 1949): 121.

16. Blake, *Public Health*, 97, 114

17. John B. Blake, "Smallpox Inoculation in Colonial Boston," *Journal of the History of Medicine* (1953): 285; Blake, *Public Health*, 112–113; Wyndham Bolling Blanton, *Medicine in Virginia in the Eighteenth Century* (Richmond: Garrett & Massie, 1931), 398–99; Fenn, *Pox Americana*, 39. After a severe outbreak of smallpox in Williamsburg occasioned by careless inoculation the Virginia House of Burgesses imposed severe restrictions on the practice in 1770.

18. D. Peter MacLeod, "Microbes and Muskets: Smallpox and the Participation of the Amerindian Allies of New France in the Seven Years' War," *Ethnohistory* 39 (1992): 46; Blake, *Public Health*, 112. Colonists had experienced virulent outbreaks of smallpox during the French and Indian War and understood that the disease would likely spread again as war with England broke out.

19. Toby Gelfand, "Military Victory, Medical Stalemate: The Battle of Princeton," in *Bicentennial Programs*, 45.

20. Mary C. Gillett, *The Army Medical Department, 1775–1818* (Washington, DC: Center of Military History, 1981), 53; George Washington to the President of the Massachusetts Congress, 10 July 1775, in Force, *American Archives*, 4:2, 1623; Robert M. Hatch, *Thrust for Canada: The American Attempt on Quebec in 1775–1776* (Boston: Houghton Mifflin, 1979), 199–200. See Louis C. Duncan, *Medical Men in the American Revolution, 1775–1783* (Carlisle Barracks, PA: Medical Fields Service School, 1931), 376–77, for an annual listing of the numbers of troops furnished by the colonies by state. This data indicates that most soldiers hailed from colonies where inoculation was not regularly available, and were thus highly likely to contract the smallpox infection if exposed.

21. Blake, *Public Health*, 112.

22. Washington to Congress, 20 July 1775, in *The Writings of George Washington from the Original Manuscript Sources, 1745–1799*, ed. John C. Fitzpatrick (39 vols., Washington, DC: Government Printing Office, 1931), 3:351.

23. Robert H. Harrison to Col. Loammi Baldwin, 13 December 1775 in Force, *American Archives*, 4:4, 255; George Washington to Joseph Reed, 15 December 1775, in *Writings of George Washington*, 4:167–68; Gillette, *Army Medical Department*, 56; Joseph Ward to Artemas Ward, 6 January 1776, in Ward Family Papers, Massachusetts Historical Society.

24. Joseph M. Miller, "Vignette of Medical History: George Washington and Smallpox," *Maryland Medical Journal* 43 (1994): 457; Stark, "Immunization Saves Washington's Army," 429; Behbehani, *Smallpox Story*, 33; J. Worth Estes, "George Washington and the Doctors: Treating America's First Superhero," *Medical Heritage* 1 (January 1985): 47; Donald R. Hopkins, *Princes and Peasants: Smallpox in History* (Chicago: University of Chicago Press, 1983), 221. According to Hopkins, the *variola* virus (smallpox) can cause male infertility. As George Washington contracted the disease in 1751, and it is commonly accepted that he was sterile, this may have been another way in which smallpox affected his life and the course of American history.

25. Blake, *Public Health*, 97–98 and 126; Samuel Bixby, "Diary of Samuel Bixby," *Proceedings of the Massachusetts Historical Society* 14 (Boston, 1876), 292. Although the British army had regularly used inoculation to protect its troops since the end of the Seven Years' War in 1763, the preventive procedure did not completely eradicate smallpox within the ranks.

26. Boyle, "Journal of the Occurrences in Boston," *New England Historical and Genealogical Register* 85 (April 1931): 119.

27. Washington's General Orders, 25 March 1776 and 19 March 1776, in *Writings of George Washington*, 4:430 and 4:404.

28. James Thacher, *The American Revolution from the Commencement to the Disbanding of the American Army*, etc. (New York: American Subscription Publishing House, 1860), 44–45; Mary Beth Norton, *Liberty's Daughters: The Revolutionary Experience of American Women, 1750–1800* (1980; reprint, Ithaca, NY, and London: Cornell University Press, 1996), 201–202; Blake, "Smallpox Inoculation in Colonial Boston," 298.

29. Fenn, *Pox Americana*, 53–54; Abigail Adams to John Adams 13 July 1776, in *The Book of Abigail and John: Selected Letters of the Adams Family, 1762–1784*, ed. L. H.

Butterfield (Cambridge, MA: Harvard University Press, 1975), 144–45; Moses Morse to commanding officer, 9 July 1776, in Force, *American Archives*, 5:1, 145; Ezekiel Price, "Diary of Ezekiel Price, 1775–1776," *Proceedings of the Massachusetts Historical* Society, 7 (1863): 259–60.

30. Gillett, *Army Medical Department*, 74; George Washington General Orders, 20 May 1776, in *Writings of George Washington*, 5:63. Montressor's Island was in the East River, and is now called Randall's Island.

31. For extensive mention of smallpox, see the Haskell, Henry, and Senter journals in Kenneth Roberts, *March to Quebec: Journal of the Members of Arnold's Expedition* (New York: Doubleday, Doran, 1938). Only two accounts of the battle neglect to mention the disease: George Morris of Pennsylvania, reflecting his probable immunity due to childhood exposure or immunization; and Ephraim Squier, who served with Gen. Roger Enos and never reached Canada. See also Ammi R. Robbins, *Journal of the Rev. Ammi R. Robbins, a Chaplain in the American Army, in the Northern Campaign of 1776* (New Haven, CT: B. L. Hamlin, 1850); Lewis Beebe, "Journal of Dr. Lewis Beebe," *Pennsylvania Magazine of History and Biography* 49 (1935): 320–61.

32. Richard L. Blanco, *Physician of the Revolution: Jonathan Potts* (New York and London: Garland STPM Press, 1979), 88–89; Roberts, *March to Quebec*, 443.

33. Blanco, *Physician of the Revolution*, 88. Blanco blames the inexperienced medical staff in Canada for the failure to institute preventive measures as soon as smallpox appeared among the troops.

34. Allen French, *The First Year of the American Revolution* (New York: Octagon Books, 1968), 689n; Benedict Arnold to Silas Deane, 30 March 1776, in Force, *American Archives*, 4:5, 549–51; "Journal of Dr. Lewis Beebe," 328; Senter Journal, in Roberts, *March to Quebec*, 238.

35. General Orders before Quebec, in Force, *American Archives*, 4:5, 551.

36. Hatch, *Thrust for Canada*, 199–200. See Duncan, *Medical Men*, 376–77, for an annual listing of the numbers of troops furnished by the colonies by state. These data indicate that most soldiers hailed from colonies where inoculation was not regularly available, and were thus highly likely to contract the smallpox infection if exposed.

37. Blanco, *Physician of the Revolution*, 89–90.

38. *The Papers of Thomas Jefferson*, ed. Julian P. Boyd et al. (Princeton, NJ: Princeton University Press, 1954), 1:437. Jefferson based his comments on firsthand accounts given to him during an investigation of the events of the Canadian campaign.

39. Benedict Arnold to Silas Deane, 30 March 1776, in Force, *American Archives*, 4:5, 549–550; Hatch, *Thrust for Canada*, 158.

40. Gillett, *Army Medical Department*, 59; *Papers of Thomas Jefferson*, 1:437; Blanco, *Physician of the Revolution*, 89.

41. Charles H. Lesser, ed., *The Sinews of Independence: Monthly Strength Reports of the Continental Army* (Chicago: University of Chicago Press, 1976), 22. In February Arnold commanded a total of 1,290 soldiers.

42. French, *First Year of the American Revolution*, 689n; Benedict Arnold to Silas Deane, 30 March 1776, in Force, *American Archives*, 4:5, 549–51; Beebe Journal, 328; Senter Journal, in Roberts, 238

43. *Journals of the Continental Congress, 1774–1789*, ed. Worthington C. Ford et al. (Washington, DC: Government Printing Office, 1904–37), 4:186.

44. A series of letters over the next few days mentioned the impossibility of enforcing Thomas's interdiction against inoculation. See John Thomas to Commissioners, 17 May 1776; and William Thompson to Commissioners, 25 May 1776, in Force, *American Archives*, 4:6, 592, 594.

45. Hatch, *Thrust for Canada*, 182, 186; James A. Huston, "The Logistics of Arnold's March to Quebec," in *Military Analysis of the Revolutionary War: An Anthology*, ed. Don Higginbotham (Millwood, NY: KTO Press 1977), 117; Howard H. Peckham, *The War for Independence: A Military History* (Chicago: University of Chicago Press, 1958), 31; John Thomas to Commissioners in Canada, in Force, *American Archives*, 4:6, 592.

46. Force, *American Archives*, 4:6, 588–89; Hatch, *Thrust for Canada*, 148–51, 183–85; Bayne-Jones, *Evolution of Preventive Medicine*, 51.

47. John Trumbull, *Autobiography, Reminiscences and Letters of John Trumbull* (New York and London: B. L. Hamlen, 1841), 27 and 299; Charles Cushing to his brother, in Force, *American Archives*, 5:1, 130; Burke Davis, *George Washington and the American Revolution* (New York: Random House, 1975), 85.

48. James E. Gibson, *Dr. Bodo Otto and the Medical Background of the American Revolution* (Springfield, IL, and Baltimore, MD: Charles C. Thomas, 1937), 102; Horatio Gates to Joseph Hawley, 10 August 1776, in Force, *American Archives*, 5:1, 901.

49. Terrence C. Davies, "American Medicine During the Revolutionary Era," *Journal of the Medical Association of the State of Alabama* 6 (November 1976): 36; Extract of a letter from John Adams, 26 June 1776, in Force, *American Archives*, 4:6, 1083. The quote continues, "I could almost wish that an inoculating hospital was opened in every town in New-England," indicating that smallpox was especially feared in that region, and that inoculation was still controversial for its residents.

50. John Sullivan to Philip Schuyler, 24 June 1776, in Force, *American Archives*, 4:6, 1201.

51. *Journals of the Continental Congress*, 5: 474, 618. See also Miller, "Vignette of Medical History," 458, where the author estimates a 40 percent mortality rate for the Canadian campaign. The Congressional Commission was originally sent to Canada for diplomatic reasons, but its attention was drawn to strategic issues and the deplorable condition of the Northern Army due to smallpox, dearth of supplies, and lack of effective leadership.

52. Washington to Jonathan Trumbull, 7 July 1776, in *Writings of George Washington*.

53. Force, *American Archives*, 4:6, 1268, Horatio Gates to the President of Congress, 2 September 1776; and 5:2, 110, Philip Schuyler to Horatio Gates, 1 September 1776.

54. George Washington to Dr. William J. Shippen, 6 January 1777; Washington to Nicholas Cooke, 10 February 1777, in *Writings of George Washington*, 6:473–74, 7:131; Fenn, *Pox Americana*, 94; Gillett, *Army Medical Department*, 75.

55. Washington to Robert Hanson Harrison, 20 January 1777, in *Writings of George Washington*, 7:38.

56. Gillett, *Army Medical Department*, 75. Washington's gamble paid off. Only four of every five hundred soldiers inoculated at Morristown in 1777 died, and Washington had all of his nonimmune troops inoculated during the winter of 1778 at Valley Forge.

57. George Washington to William Shippen Jr., 6 January 1777, in *Writings of George Washington*, 6:473. Dr. Shippen was appointed director general of the Army Medical Department in April 1777. Dr. John Morgan was relieved of that position in January 1777, so at the time of the extensive troop inoculations the Medical Department was in a state of flux and without a director general.

58. George Washington to the President of Congress, 5 February 1777, in *Writings of George Washington*, 7:102–106.

59. Fenn, *Pox Americana*, 94; Gillett, *Army Medical Department*, 74–75; William Owen, *The Medical Department of the United States Army (Legislative and Administrative History) during the Period of the Revolution, 1776–1786* (New York: Paul Hoeber, 1920), 57–58.

60. Duncan, *Medical Men*, 17; Owen, *Medical Department of the United States Army*, 57–58.

61. Stark. "Immunization Saves Washington's Army," 430–43; Andrew M. Sherman, *Historic Morristown, New Jersey: the Story of Its First Century* (Morristown, NJ: Howard, 1905), 209.

62. Stark. "Immunization Saves Washington's Army," 430–31; George Washington to William Maxwell, 18 February 1777, in *Writings of George Washington*, 7:158.

63. Stark. "Immunization Saves Washington's Army," 430–43; Duncan, *Medical Men*, 162–63, 181. Duncan reported that in 1777 smallpox spread to the civilian population in New York, hundreds of residents died, and the wealthy fled to their country homes to undergo inoculation. See also Sherman, *Historic Morristown*, 208, 217, 254–55. Sherman indicated that sixty-eight of Johnes's parishioners died of smallpox in 1777, with additional deaths in a nearby Baptist parish. The epidemic "raged uncontrolled among the inhabitants of Morristown" for nearly six months, and finally subsided by 1 September (Gelfand, "Military Victory," 47–49). In addition, a Princeton resident described how smallpox "hath got into the Neighborhood the natural way and proved very mortal both to the inhabitants and the soldiers" during the spring of 1777.

64. Richard L. Blanco, "Medicine in the Continental Army, 1775–1781," *Bulletin of the New York Academy of Medicine* 57 (1981): 687; Fenn, *Pox Americana*, 94–95; Bayne-Jones, *Evolution of Preventive Medicine*, 53; Miller, "Vignette of Medical History," 457; Blake, *Public Health*, 129.

65. Thomas Dimsdale, *The Present Method of Inoculation for the Small-Pox* (Dublin: John Exshaw, 1774). Dr. John Morgan, director general of the Army Medical Department in 1776, recommended this method; see John Morgan, *A Recommendation of Inoculation, According to Baron Dimsdale's Method* (Boston: J. Gill, 1776). For a description of the Suttonian method, see David Van Zwanenberg, "The Suttons and the Business of Inoculation," *Medical History* 22 (1978): 71–82.

66. George Washington to John Augustine Washington, 5 August 1777, in *Writings of George Washington*, 3:46, 9:21. Considerable knowledge of medicine was common among plantation owners, who had a significant investment in the health of their plantation slaves. At his death, George Washington owned and was responsible for the well-being of 216 slaves and 51,000 acres of land. Notes among Washington's papers detail the costs of inoculation for his slaves.

67. Miller, "Vignette of Medical History," 458; See the chart and graph in Lesser, *The Sinews of Independence*, xxx–xxxi for statistics on the total number of rank-and-file

soldiers. The number of soldiers available for duty fluctuated, but descended to a low of 2,188 in April 1776 at the height of the smallpox epidemic in Canada. By October of 1777 returns indicated more than 33,000 soldiers on active duty with only 17.7 percent reported ill.

68. Jones, "Medicine in Virginia," 258; Frances Norton Mason, ed., *John Norton and Sons, Merchants of London and Virginia: Being the Papers from Their Counting House for the Years 1750–1795* (Richmond, VA: Deitz Press, 1937), 110, 351.

69. Blake, *Public Health*, 131.

70. Quoted in Bayne-Jones, *Evolution of Preventive Medicine*, 54. See also Dimsdale, *Present Method of Inoculation*.

71. For a discussion of the development of the cowpox vaccine, see Derrick Baxby, *Jenner's Smallpox Vaccine: The Riddle of Vaccinia Virus and its Origin* (London: Heinemann Educational Books, 1981), 197–209; Donald A. Henderson, "The Miracle of Vaccination," *Notes and Records of the Royal Society of London* 51 (July 1997): 236. In 1796 Dr. Jenner discovered that vaccination with cowpox, an animal disease not fatal to humans, prevented the development of smallpox. Administration of this vaccine eliminated the dangerous contagion produced by inoculation. Within five years, the cowpox vaccine was available to countries on every continent, and dramatically reduced the risk of dying from smallpox.

3

The Role of the Army and Navy in American Natural History, 1803–1860

Keir B. Sterling

NATURAL HISTORY IN EARLY NINETEENTH-CENTURY AMERICA

In 1807, when Meriwether Lewis and William Clark returned from twenty-nine months in the West, American natural science was in its infancy. Few American colleges offered scientific instruction. The Military Academy at West Point, founded in 1803, was America's first engineering school. Rensselaer Polytechnic Institute in Troy, New York, opened in 1824 as the first nonmilitary scientific college. But at least through the 1850s, Philadelphia remained the "nation's foremost center" for natural history, and its scientific libraries were the country's best. The city boasted two botanical gardens—one dating from 1728, the other from 1747—and the nation's first important natural history museum, founded in 1784 by Charles Willson Peale. It was home to the American Philosophical Society, founded in 1743 as America's first scholarly organization, and to the University of Pennsylvania, which began medical instruction in 1765. Early in the nineteenth century, the American Botanical Society, founded in 1806, soon became the Linnean Society and then in 1812 metamorphosed into the Academy of Natural Sciences, which added to its collections of geological and "natural objects" a library in 1816 and a journal in 1817.[1]

Elsewhere, natural science was slower to develop. Some New England colleges began offering instruction in geology and botany soon after 1800. In 1830 Massachusetts initiated the first of several dozen state geological and natural history surveys. Louis Agassiz's pioneering program in zoology at Harvard got under way in 1848.[2] But "science languished" at the U.S. Naval Academy, which opened in 1845—a full forty years after West Point.

Many naval officers disdained science and scientists, particularly those outside the military.[3] In Washington, the Smithsonian Institution opened its doors in 1846. Soon afterward, the redoubtable Spencer Fullerton Baird began his thirty-seven-year tenure, busily assembling legendary natural history collections as the Smithsonian's first assistant secretary. Yet it would take another decade for the U.S. National Museum to come into being. And not until after the Civil War did professional and institutional science reach the stage where American colleges could launch postgraduate training programs.[4]

Before 1850, therefore, American army and navy explorers had a limited knowledge base with which to work. Few European and American articles or books on American natural history had been published before the American Revolution. In the early nineteenth century, their number and sophistication increased, but few American naturalists published their original findings before 1830. Neither European nor American scientific literature was readily available outside Philadelphia. Before going into the field, military men could consult only a handful of civilian naturalists. Among those known to have provided information to Lewis and Clark and later army explorers through their books, all published in Philadelphia (and in person), were Benjamin Rush, *Medical Inquiries and Observations*, (5 vols., 1789–1798); Charles Willson Peale, *Introduction to a Course of Lectures on Natural History Delivered in the University of Pennsylvania* (1799); Benjamin Smith Barton, *Elements of Botany* (1803); and Caspar Wistar, *System of Anatomy* (1811).[5]

Making preliminary research even more difficult was the scarcity of plant and animal specimens available for study. Owing to primitive early nineteenth-century methods of preserving animal specimens, only a few skins have survived to the present day. For example, less than a handful of Lewis and Clark's animal specimens, many of which had been sent to C. W. Peale, still exist. Peale died in 1827 and his museum closed in 1848. His collections were sold or lost, owing to fire and other mishaps. Fortunately, the plants collected by Lewis, Clark, and later military explorers fared somewhat better.[6]

Despite this lack of background knowledge, beginning in 1803 the military launched an impressive number of scientific expeditions. Justifications abounded for both the army and navy to sponsor these expeditions. The army stressed settlement of new lands, as well as the inventorying and exploitation of natural resources. The navy emphasized commerce, whaling, and oceanography. When foreign nations competed for new territories, as in the Southwest and the Mississippi River Basin, those countries possessing more comprehensive understanding of an area's climate, geography, and fauna and flora often took possession. So knowledge often led to political and military advantage. Finally, many Americans were influenced by a Humboldtean worldview, an early nineteenth-century intellectual variety of "manifest destiny," which held that educated persons should have some knowledge of all or most subjects.[7]

Why were relatively junior military officers chosen to lead early exploring expeditions? Most were younger men (the ages of the four discussed here ranged from twenty-nine to forty-three). All were physically fit and trained to be leaders. The army and navy were small and had few wars to fight. Promotion was based strictly upon seniority rather than merit, so advancement was glacially slow. Given these conditions, many young officers were eager to escape monotonous garrison duty for adventurous exploration in forests, mountains, and deserts. Their careers are not commonly remembered, but these officers were adventurous and productive.[8]

HOWARD STANSBURY (1806–1863)

Howard Stansbury, a New York native, began his career as an enterprising civil engineer, surveying projected canal and railroad routes rights of way in the midwestern states and territories. He had charge of public works projects in Indiana and in Richmond, Virginia. In 1838 he accepted a commission as lieutenant in the newly formed Army Corps of Topographical Engineers, and was promoted to captain two years later, at age thirty-four. The army continually employed a modest number of topographical engineers from the 1770s on, but the Corps of Topographical Engineers was in existence only from 1838 until 1863, when it was absorbed into the Army's Corps of Engineers.[9]

Stansbury spent years making surveys of the Great Lakes and the harbor of Portsmouth, New Hampshire. He is best known, however, for his march into the Great Salt Lake region with an eighteen-man military party in 1849 and 1850. Although not the first to reach that area—Capt. Benjamin Bonneville had visited the Salt Lake in 1831, Capt. John C. Fremont in 1843–44—Stansbury's party amassed the first detailed study of the Salt Lake region. With canny foresight, Stansbury persuaded the suspicious Mormon leader Brigham Young to cooperate with his survey. Clearly, Stansbury genuinely respected what the Mormons had accomplished, and contemporary readers were intrigued with what he had to say about Mormon society. Stansbury also wrote of encounters with Utah's unusual fauna and fossil vertebrates.[10]

Stansbury's report, *Exploration and Survey of the Valley of the Great Salt Lake of Utah, Including a Reconnaissance of a New Route through the Rocky Mountains,* included scientific appendices on the animals, plants, and fossils collected, some of them new to science. The appendixes were prepared in Washington by contemporary authorities such as Spencer Baird of the Smithsonian, Titian

Peale, the French-born zoologist Charles Girard (1822–1895), the botanist John Torrey (1796–1873); the geologist and paleontologist James Hall (1811–1898), and Samuel S. Haldeman (1812–1880), a natural historian, chemist, and philologist.[11]

Published as a U.S. Senate Document in 1852, Stansbury's report went through several commercial editions and is still in print today. This document comprised the very first published account of the geography, geology, plant, and animal resources of the region.[12] In later years, Stansbury made other surveys and built military railroads in Minnesota. Promoted to major, he died while serving as a mustering and disbursing officer in Ohio and Wisconsin during the Civil War.[13]

JOSEPH CHRISTMAS IVES (1828–1868)

Joseph Christmas Ives, born in New York City, was raised in modest circumstances in a New Haven, Connecticut, boarding house. After attending Yale, he completed his studies at West Point in 1852. Initially assigned to ordnance duties, he soon transferred to the Topographical Engineers and joined Lt. Amiel Whipple's Pacific Railway Survey expedition of 1853–1854, which traveled to California through Arkansas and the New Mexico and Arizona Territories. Returning to Washington, Ives spent three years in the survey's home office.[14]

In 1857 the army instructed Ives to find ways of supporting, by land or water, the troops fighting Mormons in Utah. Ives and his party bought a fifty-four-foot iron-hulled steamboat in Philadelphia, which they dubbed *Explorer*. The disassembled craft was shipped by steamer via Panama to California and then back to a point near Yuma, Arizona. There, Ives and his men reassembled the vessel, and in late December 1857 they steamed up the Colorado River. A short time later, *Explorer* hit a rock, damaging the steamboat and upsetting Ives's plans. There was nothing for it but to send one group of men back to Yuma with *Explorer*. In the meantime, Ives and about thirty others slowly continued along the banks of the river by land.[15]

In early April 1858 they stumbled upon an astonishing experience. The party hiked into the floor of the Grand Canyon, probably the first U.S. citizens to do so. Native Americans, of course, knew the region and Spanish explorers also had discovered the canyon in 1540. In any event, while still at the Canyon's base, John Strong Newberry, the expedition doctor, naturalist, and geologist, recorded the stratified rock formations on the canyon walls. Heinrich Möllhausen, the expedition artist-naturalist, obliged by drawing colored illustrations of the rock layers. This geological "measuring stick," plus the unique relief map of the West prepared by expedition topographer F. W. von Egloffstein, became important foundation works for many later geologists and geographers working in the West.[16]

Ives subsequently pushed into southern Utah and east to Fort Defiance, Arizona, before returning to Yuma in May 1858. His naturalists returned East by land. Ives's *Report upon the Colorado River of the West*, published in 1861, was considered a work of great scientific and literary merit, and von Egloffstein's map was outstanding. In two appendices, Spencer Baird of the Smithsonian prepared a list of the fifty-seven birds collected by Möllhausen, and four botanists compiled a twenty-five-page account of the plants collected.[17] In his journal, Ives remarked that the Grand Canyon "look[ed] like the Gates of Hell." Looking at the region through the eyes of an engineer, he thought it "altogether valueless," and felt that it would remain "forever unvisited and undisturbed." Ives expected to "doubtless be the last . . . to visit this profitless locality."[18]

Returning to Washington, Ives directed work on the Washington Monument during the years 1859 and 1860. During the Civil War, he joined the Confederate army as a captain of engineers. His wife was a Southerner, and Ives greatly respected Col. R. E. Lee and former secretary of war Jefferson Davis, both of whom he had known before 1861. Ives rose to the rank of colonel, and became an aide-de-camp to Confederate president Jefferson Davis. After the war, Ives moved with his family back to his native New York City, where he died at the early age of forty.[19]

CHARLES WILKES (1798–1877)

Beginning in the 1830s, many naval officers did notable scientific exploration. Collectively, they helped disprove a claim made to several congressmen in 1831, that "there is not an officer of the Navy capable to conduct a scientific enterprise."[20] New York–born Charles Wilkes was educated at home by his father. He joined the merchant service at seventeen and became a naval midshipman when he was twenty. In 1832 and 1833, as a lieutenant, he surveyed Narragansett Bay and then began a five-year tour as officer in charge of the Navy's Department of Charts and Measurements.[21]

In 1842 Wilkes took command of the U.S. Exploring Expedition, for which Congress had appropriated $330,000. The expedition sailed on six unsuitable and poorly equipped ships with 82 officers, 342 sailors, and 9 civilian scientists. The principal objectives were to chart the Pacific for navigation, "to extend the bounds of science and promote knowledge," and to find new hunting grounds for American sealers. Lacking sea experience, Wilkes got the assignment only after four more senior officers had either declined the command or resigned. He nonetheless performed most duties well. In four years, his ships sailed around South America and along America's Pacific Coast, mapped fifteen hundred miles of Antarctic shoreline, and crisscrossed the Pacific. The expedition covered eighty-seven thousand miles, surveyed 280 Pacific islands, and circled the globe, all at an ultimate cost of $900,000—a huge price tag at that time, but cheap for the knowledge acquired.[22]

The expedition's cost included men and ships as well as money. Only two of the original fleet of six came home. Somewhere off the southern coast of Latin America, one ship disappeared with all hands, never again to be heard from. Wilkes sent the store ship *Relief* home both because it sailed too slowly and because he wished to be rid of a group of officers and men with whom he had clashed or with whose services he was dissatisfied. While in Australia, one of the scientifics, Joseph Pitty Couthoy, a conchologist, became ill and went to Hawaii for his health. Wilkes, with whom Couthoy had quarreled, refused to allow him to rejoin the expedition and sent him home on another vessel with a young officer Wilkes had dismissed, a sick midshipman, and some plants collected by the expedition in Fiji.[23]

Wilkes sold a third ship, which had become unserviceable, in Singapore. According to local rumor, the new British owner of the *Flying Fish* intended to use it "as an opium smuggler."[24] Three years into the voyage, in July 1841, a fourth expedition vessel, USS *Peacock*, foundered off the Oregon coast, and a replacement had to be purchased. During the four-year trip, 178 new men joined the expedition at various ports of call, including Honolulu, Sydney, and Rio de Janeiro. Of all the men ultimately involved in the enterprise, 24 died from various causes or were lost at sea, 95 others were either discharged as unsuitable or returned on other vessels, and 127 deserted—a testament to the sometimes harsh shipboard conditions. One of the dead was Wilkes's nephew, killed by natives in Fiji. Of the original complement of men, only 181 returned home with the expedition.[25] Brought back as a captive was Veidovi, a Fijian chief who had murdered American sailors visiting his island in 1834. Veidovi died following the expedition's arrival in the United States. His head was removed and placed in spirits, and "his fleshless skull became a permanent part of the Expedition's collection."[26]

Not everything collected reached home. Titian Peale deemed many of the sailors "perfect children," with little understanding of the significance of scientific specimens.[27] On one occasion, a sailor tossed some valuable specimens overboard because they smelled bad.[28] Peale personally brought back 2,150 bird specimens, 134 mammals, 588 fish, and some 40 species of Pacific Northwest butterflies, but he also lost some 350 to 400 butterfly species collected on Pacific islands, together with other materials and notes, when the *Peacock*, on which he was billeted, foundered off the Oregon coast.[29] Far greater losses followed the secretary of the navy's unfortunate decision to have sometimes-incompetent political appointees open and sort shipments from Wilkes before the expedition's return. A number of specimens preserved in alcohol dried out, and many others lost their labels or identifying documents. In at least one instance, a male and a female bird were in some fashion sewn together as one. Peale also claimed that these same incompetents lost 180 of his bird specimens, some of them new species.[30]

Nevertheless, the scientists returned to Washington with forty tons of valuable plant and animal specimens, maps, charts, anthropological artifacts and meteorological data. Peale and the other scientists also brought back 300 species of fossil animals, 400 species of coral, 1,000 species of crustaceans, 200 jars of insects, and other preserved zoological specimens. The botanical collectors returned with 50,000 specimens of plants, representing 10,000 species, and seeds for propagating 648 species.[31] One nautical chart was so well done that one hundred years in the future, U.S. Marines would depend upon it for the invasion of Tarawa in 1943.[32]

The volume of material eventually helped create the U.S. National Museum, the U.S. Botanic Garden, and the National Herbarium.[33] The continuing acquisition of specimens from the Pacific Railway Surveys of the 1850s and other western expeditions following the Civil War ensured the National Museum's continued existence. Spencer F. Baird, as the Smithsonian's assistant secretary from 1850 to 1878, and as second secretary of the institution from 1878 until his death in 1887, helped to "tilt the Smithsonian's progress toward the biological sciences and away from the physical."[34] Wilkes's nephew, James Renwick Jr., became the architect for the Smithsonian's "Castle" on the Mall in Washington.[35] The growth of the National Museum enhanced the city's standing as a center of scientific research. Before the mid-nineteenth century, only 9 percent of American scientists had spent "any part of their working lives" in Washington.[36]

Of the twenty-five projected descriptive Exploring Expedition volumes and ten atlases (atlases in this case contained pictures of various animals, plants, other materials collected, and charts), four volumes of text, one atlas, and two volumes of charts were never published, one volume was suppressed, and one other was never officially distributed. The total cost of publication from 1844 to 1874 was $348,798.16. The civilian contributors and artists received $76,432.28 for their work and the expenses they incurred on thirteen of the volumes, the ten atlases, and the two volumes of charts in the series, including the four volumes that were never published and the one (by Peale) which was suppressed. This does not include Wilkes's navy pay, though he wrote a five-volume *Narrative of the Expedition* and three volumes of meteorological and hydrographic observations, one of which remained unpublished. The final cost of the expedition itself was $928,183.62.[37]

Some U.S. senators objected to spending money on scientific publications. Sen. Simon Cameron (R-OH) asked, "What do we care about stuffed snakes, alligators, and all such things?" and added, "I am tired of this thing called science here."[38] Sen. John P. Hale (R-NH) objected to printing "pictures of bugs, snakes, and reptiles," and to "spend[ing] $10,000 to distribute them . . . the thing is all wrong, Sir." But at least one (unnamed) congressman wanted to have a species named after him.[39] Foreign and domestic distribution of Exploring Expedition volumes was handled by the U.S. Department of State. Congress shortsightedly authorized only one hundred copies of each published volume. Many of the sets distributed to the individual states were lost to fire, destroyed during the Civil War, or simply could no longer be found in the 1930s and early 1940s. Other volumes went unpublished owing to congressional parsimony, notably, studies by Louis Agassiz on fish and by Asa Gray on plants. However, Wilkes's *Narrative* and several of the scientific reports are still in print.[40]

Until 1874, when Congress finally terminated the project, Wilkes oversaw publication of eighteen scientific tomes other than his own based on the expedition's material. The writers, expedition scientists and other authorities, often found Wilkes's oversight trying. He would not, for instance, permit naturalists working with their specimens in Washington either to compare them with museum specimens in Philadelphia or to consult needed books in Philadelphia libraries. He also wanted all scientific Latin translated into English to benefit the general public. His relations with Titian Peale were particularly strained. Specifically assigned responsibility for collecting mammals and birds prior to the expedition's departure, Peale's volume, provisionally entitled *Zoology*, was never published. Wilkes suppressed the volume on grounds that it was scientifically unsound, although many modern zoologists disagree.[41] He had the work redone by John Cassin, a Philadelphia naturalist who had not accompanied the expedition and who used much of Peale's material and illustrations. Many of Peale's plates appeared in Cassin's 1858 publication.[42]

Wilkes was not Peale's only problem. He faced continuing personal and professional difficulties while working on his volume. Within a three-year period (1844–1847), his wife and two of his children died, leaving him with two teenaged children to care for. The family museum in Philadelphia failed, there was a sheriff's sale, and the collections there were scattered. Peale then unsuccessfully sought a curatorial position with the new Smithsonian Institution. Both Wilkes and Peale were proud and difficult men who frequently disagreed about Peale's role in the overall publication scheme. Wilkes discharged Peale from his salaried position as author and artist after suppressing his book, but offered to pay him twenty dollars for each additional illustration he completed. Thoroughly fed up with the manner in which Wilkes had handled matters, Peale washed his hands of the project and spent the rest of his working life as an examiner for the U.S. Patent Office.[43] Stanton notes "the passing of the old-fashioned naturalist and [the] summoning [in] his stead [of] the coolly proficient writer of technical monographs, who addressed himself not to nature but to atoms and cells. Peale never fully grasped the implications of what had happened when art and science went their separate ways."[44]

Unfortunately, despite the very real accomplishments of this intelligent and determined officer, the expedition was marked by controversy. Wilkes was a tactless, single-minded disciplinarian who had antagonized most of his officers, crew, and his nine "scientifics." A relentless faultfinder throughout the expedition's four years at sea, he disciplined and arrested several officers during the voyage, and was himself court-martialed upon his return to the United States. In proceedings held aboard the USS *North Carolina* in New York Harbor in the summer and early fall of 1842, Wilkes brought charges against four junior officers "and they against him." Following five trials, which ran from July into September, one man, a surgeon, was sentenced to be dismissed from the service, a sentence President John Tyler commuted to one year's suspension from duty. Two others were reprimanded, one also receiving a six-month suspension. The fourth man was acquitted. Wilkes himself was found guilty of "seventeen instances of illegal punishment," principally involving the whipping of enlisted personnel. The furious Wilkes, who had anticipated promotion for the expedition's accomplishments, was instead reprimanded by the secretary of the navy. Although promoted to commander in July 1843, Wilkes had to wait until 1856 for his much-coveted captaincy.[45]

Wilkes's subsequent naval career continued on a noteworthy path. During the Civil War, he forcibly removed two Confederate commissioners from a British vessel and, because of this unauthorized capture, nearly precipitated war with England. Following several other high-handed actions, during which he had commandeered several vessels of his own and other nations, Wilkes was court-martialed in the summer of 1863, publicly reprimanded by the secretary of the navy, and suspended from duty for three years. He retired in 1866 as a rear admiral. Wilkes's "naval career had spanned half a century, but he had never in the course of it learned to be a subordinate." He died in 1877.[46]

WILLIAM LEWIS HERNDON (1813–1857)

Naval officer William Lewis Herndon led a less dramatic, but nonetheless valuable scientific expedition to South America. Born in Fredericksburg, Virginia, in 1813 and orphaned when very young, Herndon entered the navy as a midshipman at age fifteen and was promoted to lieutenant in 1841, when he was twenty-eight. Herndon was both a younger cousin and brother-in-law of Matthew Fontaine Maury, the pioneering American naval officer and oceanographer. After working for Maury in the Navy's Department of Charts and Measurements for several years, Herndon commanded a navy steamer during the Mexican War. After the war, he again worked for Maury at the Naval Observatory, and then assumed command of the USS *Vandalia*.

When the *Vandalia* docked at Santiago, Chile, in 1851, Herndon found new instructions awaiting him from Maury. Spurred by earlier British explorations of the Amazon Basin, Maury ordered Herndon to investigate the navigability of the Amazon River, its geography, and its suitability for agriculture and commerce. Items of scientific interest were also to be collected. In confidence, Maury told his cousin that American slaves might some day be resettled in the Amazon region, and that American planters might profitably be engaged in agriculture there. Among his other qualifications, Herndon "could translate Latin, Spanish, and Portuguese."[47]

Maury left the choice of exploratory routes up to Herndon, who spent months organizing and gathering supplies in Santiago and Lima, Peru, for his foray into Amazonia. Hoping for some useful specimens, the Smithsonian's Spencer Baird almost certainly sent specimen-collecting instructions and supplies to Herndon. Altogether, Herndon spent a year and traveled four thousand miles across the Andes and down the Amazon to its mouth. In addition to geographic and navigational data, he collected local animals, plants, and seeds. Some of the animals he skinned, while others were taken along in cages. Unfortunately, the animals often fought or killed each other. When Herndon left Brazil, he still had a tortoise, perhaps a dozen pairs of six species of birds, and several mammals. Sadly, his last caged monkey expired in July 1852, as his ship entered New York Harbor.[48]

Thanks to the noted English naturalist Henry Walter Bates (1825–1892), a beetle and butterfly specialist, Herndon also brought back an array of butterfly specimens. Herndon met Bates in 1852 near the town of Santarem in Brazil. Bates had been exploring the Amazon region since 1845 and would remain until 1857. He collected more than fourteen thousand insects and other animals, eight thousand of them new to science. His 1863 account, *Naturalist on the River Amazons*, has become a classic.[49]

In a side trip, Herndon's assistant Leander Gibbon had explored the great river's Bolivian tributaries with a separate group of men. He returned eight months after Herndon with additional material. The Smithsonian got both Herndon's and Gibbon's specimens. Although the government could not find the money to print a catalog of their collection, they were able to publish their two-volume report with maps. Gibbon's report was pedestrian, but Herndon's account was so impressive that the government printed thirty thousand copies. When the young Samuel Clemens (Mark Twain, 1835–1910) read it, he was inspired to leave his home in Hannibal, Missouri, for New Orleans in hopes of taking passage to Brazil. Clemens failed in that attempt, but would later note the influence Herndon's account had on his later writing career.[50] Explorers and anthropologists mined it for years, and the volume remains in print today.[51]

After several years of sea duty, Herndon took a leave of absence from the navy to skipper merchant ships. Sailing from Panama to New York in September 1857, Herndon's ship, the *Central America*, went down in a gale off Cape Hatteras, North Carolina. Passing vessels managed to save 149 passengers, mostly women and children, but Herndon, 425 other passengers, and many gold bars from California went down in America's worst maritime disaster. A monument to Herndon's memory at the Naval Academy has become part of a modern spring ritual, as midshipmen celebrating the end of their plebe year climb to its top. In an interesting afterward to the story, Herndon's orphaned daughter, Ellen, later married a young New York attorney named Chester Alan Arthur. Always frail, she died of pneumonia in 1880 at age forty-two, the year before her husband became twenty-first president of the United States.[52]

THE SIGNIFICANCE OF MILITARY EXPLORATION

The collective efforts of these four officers—Stansbury, Ives, Wilkes, and Herndon—gave a considerable boost to American natural science in its formative years. These four are representative of hundreds of young military men in the nineteenth century who adventured their way into scientific and exploratory history. We remain indebted to them today.

NOTES

1. Frederick Rudolph, *The American College and University: A History* (New York: Vintage, 1965), 228–31; Robert V. Bruce, *The Launching of Modern American Science* (Ithaca, NY: Cornell University Press, 1988), 130, 161; Brooke Hindle, *The Pursuit of Science in Revolutionary America, 1735–1789* (Chapel Hill: University of North Carolina Press, 1956), 20–28, 136–40, 261–62, 308, 309, 316, 320–21, 323, 325; Herman Viola and Carolyn Margolis, eds., *Magnificent Voyagers: The U.S. Exploring Expedition, 1838–1842* (Washington, DC: Smithsonian Institution Press, 1985), 233; Dirk J. Struik, *Yankee Science in the Making*, rev. ed. (New York: Collier Books, 1962), 43; William M. Smallwood and Mabel S. C. Smallwood, *Natural History and the American Mind* (New York: Columbia University Press, 1941), 130–42.

2. Struik, *Yankee Science*, 219–34; Rudolph, *American College and University*, 223–28; 231–232; Bruce, *Launching of Modern American Science*, 164; Samuel Eliot Morison, *Three Centuries of Harvard, 1636–1936* (Cambridge, MA: Harvard University Press, 1936), 279.

3. Bruce, *Launching of Modern American Science*, 161.

4. Bruce, *Launching of Modern American Science*, 187–88, 198–99, 232, 233, 326, 334–38; K. B. Sterling, "Zoological Research, Wildlife Management, and the Federal Government," in *Forest and Wildlife Science in America: A History*, ed. H. K. Steen (Durham, NC: Forest History Society, 1999), 22–23; A. Hunter Dupree, *Science in*

the Federal Government: A History of Policies and Activities (Baltimore, MD: Johns Hopkins University Press, 1986), 45, 46; Rudolph, *American College and University*, 335, 338, 339–43, 348–49.

5. Smallwood and Smallwood, *Natural History and the American Mind*, 35–40, 125–28, 140–42, 185–86, 188; K. B. Sterling, "Lewis, Meriwether," in *Biographical Dictionary of American and Canadian Naturalists and Environmentalists*, ed. K. B. Sterling et al. (Westport, CT: Greenwood Press, 1997), 465.

6. Judith Chupasko, Museum of Comparative Zoology, Harvard University, personal correspondence with author, 29 August 2003; Sterling, "Lewis, Meriwether," 465, 466; Sterling, "Zoological Research," 20.

7. Elizabeth Green Musselman, "Science as a Landed Activity: Scientifics and Seamen aboard the U.S. Exploring Expedition," in *Surveying the Records: North American Scientific Exploration to 1930*, ed. Edward C. Carter III (Philadelphia, PA: American Philosophical Society, 1999), 80; Bruce, *Launching of Modern American Science*, 183, 206.

8. Edward M. Coffman, *The Old Army: A Portrait of the American Army in Peacetime, 1784–1898* (New York: Oxford University Press, 1986), 49–50, 56, 60, 67; Francis P. Prucha, *The Sword of the Republic: The United States Army on the Frontier, 1783–1846* (Bloomington: Indiana University Press, 1977), 144–47, 162–63.

9. See Frank N. Schubert, ed., *The Nation Builders: A Sesquicentennial History of the Corps of Topographical Engineers, 1838–1863* (Fort Belvoir, VA: Office of History, Army Corps of Engineers, 1988).

10. Joseph Ewan and Nesta Dunn Ewan, *Biographical Dictionary of Rocky Mountain Naturalists: A Guide to the Writings and Collections of Botanists, Zoologists, Geologists, Artists, and Photographers, 1682–1932* (Utrecht and Antwerp: Bohn, Scheltema & Holkema; The Hague and Boston: Dr. W. Junk, 1981), 210; Vasco M. Tanner, "A Chapter on the Natural History of the Great Basin, 1800–1855," *Great Basin Naturalist* 1 (1940): 33–61; "Stansbury, Howard," in *Encyclopedia of Frontier Biography*, ed. Dan L. Thrapp (Lincoln: University of Nebraska Press; Spokane, WA: Arthur H. Clark, 1988), 1356; K. B. Sterling, "Stansbury, Edward," in *Biographical Dictionary*, ed. Sterling et al., 743–45; Francis B. Heitman, *Historical Register and Dictionary of the United States Army, from Its Organization, September 29, 1789, to March 2, 1903* (Washington, DC: Government Printing Office, 1903), 915.

11. Howard Stansbury, *Exploration and Survey of the Valley of the Great Salt Lake of Utah, Including a Reconnaissance of a New Route through the Rocky Mountains*, U.S. Sen., Special Sess., March 1851, Exec. Doc. No. 3 (Philadelphia, PA: Lippincott, Grambo, 1852), 4–5, 84–86, 130–136, 138–140. Appendices: "Mammals," by Spencer F. Baird, 309–13; "Birds," by Spencer F. Baird, 314–35; "Reptiles," by Spencer F. Baird and Charles Girard, 336–65; "Insects," by Samuel S. Haldeman, 366–79; "On Certain Insect Larvae," by Titian R. Peale, 379; "Botany," by John Torrey, 383–97; Letter from James Hall on Geology and Paleontology, 401–14. Online versions of many reports of expeditions made by the Topographical Engineers, including the two discussed in this paper, are at http://www.topogs.org/Homepage.htm

12. L. J. Dorr, D. H. Nicholson, and L. K. Overstreet, "Bibliographic Notes on Howard Stansbury's Exploration and Survey of . . . Expedition to the Valley of the Great Salt Lake," *Archives of Natural History* 30, pt. 2 (October 2003): 317–30. There were three slightly different versions of the first edition of 1852, one of them

reprinted in 1855, and a second edition in 1853, together with an 1854 German translation. In the second edition, the meteorological appendix was expanded by eight pages, and there were changes of nomenclature in the botanical appendix. Several plates were misnumbered in the first edition. Modern reprints are of the first edition. See also Max Meisel, "Stansbury Expedition," in his *A Bibliography of Natural History: The Pioneer Century, 1769–1865* (Brooklyn, NY: Pioneer Publishing, 1929), 3:114–16. Heitman, *Historical Register*, 565.

13. William Goetzmann, *Army Exploration in the American West, 1803–1863* (New Haven, CT: Yale University Press, 1959), 219–25; Goetzmann, *Exploration and Empire: The Explorer and Scientist in the Winning of the American West* (New York: Alfred A. Knopf, 1966), 277–79, 432; Frank N. Schubert, *Vanguard of Expansion: Army Engineers in the Trans-Mississippi West, 1819–1879* (Washington, DC: Office of the Chief of Engineers, 1980), 83–86.

14. Dan L. Thrapp, "Ives, Joseph Christmas," in *Encyclopedia of Frontier Biography*, ed. Thrapp, 710–11; Goetzmann, *Army Exploration*, 378–79, 380–94.

15. Goetzmann, *Exploration and Empire*, 306–308, 316; Max Meisel, "Ives Expedition," in *Bibliography of Natural History*, 3:271–73; Donald Worster, *A River Running West: The Life of John Wesley Powell* (New York: Oxford University Press, 2001), 129–31; Schubert, *Nation Builders*, 86–90.

16. Kathleen Doherty, "Möllhausen, Heinrich Balduin," *The Handbook of Texas Online*, http://www.tsha.utexas.edu/handbook/online/articles/view/MM/fmbot.html; J. B. Krygier, "Envisioning the American West: Maps, the Representational Barrage of 19th Century Expedition Reports, and the Production of Scientific Knowledge," *Cartography and GIS* 24 (1997): 27–50; Ben W. Huseman, *Wild River, Timeless Canyon: Balduin Möllhausen's Watercolors of the Colorado* (Tucson: University of Arizona Press, 1995).

17. Joseph C. Ives, *Report upon the Colorado River of the West, Explored in 1857 and 1858* (Washington, DC: Government Printing Office, 1861), part 3, *Geology Report*, by John Newberry (160 pp.); part 4, *Botany*, by Asa Gray, John Torrey, George Thurber, and George Englemann (32 pp.); and part 5, *Zoology* [confined to birds only], by Spencer F. Baird (8 pp.), were all separately paginated.

18. Schubert, *Nation Builders*, 61.

19. Thrapp, "Ives," 710–11; Goetzmann, *Army Exploration*, 432, 433.

20. As quoted in Steven J. Dick, *Sky and Ocean Joined: The U.S. Naval Observatory, 1830–2000* (New York: Cambridge University Press, 2003), 44.

21. Clark Elliott, *Biographical Dictionary of American Science: The Seventeenth through the Nineteenth Centuries* (Westport, CT: Greenwood Press. 1979), 274–75.

22. Daniel C. Haskell, *The United States Exploring Expedition and Its Publications, 1844–1874* (New York: Greenwood Press, 1968), 3, 5; Nathaniel Philbrick, *Sea of Glory: America's Voyage of Discovery, the U.S. Exploring Expedition, 1838–1842* (New York: Viking, 2003), xv–xix, 307–30; Musselman, "Science as a Landed Activity," 83, 86; William Stanton, *The Great United States Exploring Expedition of 1838–1842* (Berkeley and Los Angeles: University of California Press, 1966), 19–20, 29, 61; Viola and Margolis, eds., *Magnificent Voyagers*, 189–92.

23. Stanton, *Great United States Exploring Expedition*, 219, Daniel B. Tyler, *The Wilkes Expedition: The First United States Exploring Expedition (1838–1842)* (Philadelphia, PA: American Philosophical Society, 1968), 125, 197; Philbrick, *Sea of Glory*, 141, 220–21, 237.

24. Philbrick, *Sea of Glory*, 220–221, 293.

25. Musselman, "Science as a Landed Activity," 79, 91; Philbrick, *Sea of Glory*, 109, 111–12, 220–21, 270–74, 278–79, 293, 300; Viola and Margolis, *Magnificent Voyagers*, 94, 155, 156, 196, 257–65. A complete list of expedition members, together with their ranks, the ships they served on, and comments concerning the extent of their participation can be found on the Christchurch NZ City Library website at http://library.christchurch.org.nz/Heritage/USExploringExpedition/.

26. Philbrick, *Sea of Glory*, 300. See also Tyler, *Wilkes Expedition*, 164–66, 183, 230, 242, 271, 361–62, 368–70.

27. Musselman, "Science as a Landed Activity," 93.

28. Musselman, "Science as a Landed Activity," 97.

29. Haskell, *United States Exploring Expedition* 60, 130; Viola and Margolis, *Magnificent Voyagers*, 234.

30. Titian Peale, letter to George Ord, 14 March 1843, American Philosophical Society, Philadelphia, PA; Haskell, *United States Exploring Expedition*, 130; Viola and Margolis, *Magnificent Voyagers*, 234; Jessie Poesch, *Titian Ramsay Peale and His Journals of the Wilkes Expedition* (Philadelphia, PA: American Philosophical Society, 1961), 96, 409.

31. Philbrick, *Sea of Glory*, 332.

32. Robert Sherrod, *Tarawa: The Story of a Battle* (New York: Duell, Sloane, & Pearce, 1944), 19; Viola and Margolis, *Magnificent Voyagers*, 187.

33. Stanton, *Great United States Exploring Expedition*, 359; Philbrick, *Sea of Glory*, xv, xix, 351; William Goetzmann, "The Great United States Exploring Expedition and the Multi-Dimensional Frontier," in *Military Influences on Washington [State] History: Proceedings of a Conference*, ed. William Woodward and David Hansen (Tacoma, WA: Washington Army National Guard, 1984), 16.

34. Viola and Margolis, *Magnificent Voyagers*, 248–53.

35. Philbrick, *Sea of Glory*, 350.

36. Bruce, *Launching of Modern American Science*, 48.

37. Haskell, *United States Exploring Expedition*, 6, 16, 17, 18–19. For further details on Wilkes Expedition publications, see Meisel, *Bibliography of Natural History*, 2:650–73.

38. Dupree, *Science in the Federal Government*, 83.

39. Haskell, *United States Exploring Expedition*, 23–24.

40. Haskell, *United States Exploring Expedition*, 25.

41. Musselman, "Science as a Landed Activity," 85.

42. On the present evaluation of Peale's and Cassin's work, see Tyler, *Wilkes Expedition*, 411–12.

43. Poesch, *Titian Ramsay Peale*, 95, 98–99.

44. Stanton, *Great United States Exploring Expedition*, 328.

45. Tyler, *Wilkes Expedition*, 374, 375–85; Philbrick, *Sea of Glory*, 329–30, 332. Stanton, *Great United States Exploring Expedition*, 284–88

46. Philbrick, *Sea of Glory*, 349, 353–56, 357–58; Stanton, *Great United States Exploring Expedition*, 360, 363.

47. William Lewis Herndon, *Exploration of the Valley of the Amazon, 1851–1852, Made Under the Direction of the Navy Department*, part 1, 33rd Cong., 1st Sess., H. Rep. (Washington, DC: Robert Armstrong, 1854), ed. and intro. Gary Kinder (New York: Grove Press, 2000), xviii.

48. Vincent Ponko, *Ships, Seas and Scientists: U.S. Naval Exploration and Discovery in the Nineteenth Century* (Annapolis, MD: Naval Institute Press, 1974), 61–79; K. B. Sterling, "Herndon, William Lewis," in *Biographical Dictionary*, ed. Sterling et al., 366–69; Herndon/Kinder, *Exploration of the Valley of the Amazon*, 4, 6–7, 19–23, 94–97, 142–43, 149–50, 151, 154, 157, 165–66, 175, 178, 181, 191, 203–204, 206, 215, 235–36, 242–43, 245–46, 251, 252, 253, 254–55, 266.

49. Herndon/Kinder, *Exploration*, 305; Alex Shoumatoff, introduction to *The Naturalist on the River Amazons*, by Henry Walter Bates (New York: Penguin Books, 1989), xii.

50. Herndon/Kinder, *Exploration of the Valley of the Amazon*, xvi.

51. Ponko, *Ships, Seas and Scientists*, 79–92; Max Meisel, "Herndon Expedition," in *Bibliography of Natural History*, 2:126–27.

52. Ponko, *Ships, Seas and Scientists*, 92; Gary Kinder, *Ship of Gold in the Deep Blue Sea* (New York: Atlantic Monthly Press, 1998), 22, 23, 33–75, 123–36, 200–203; William A. DeGregorio, *The Complete Book of U.S. Presidents: From George Washington to George Bush* (Avenel, NJ: Wings Books, 1991), 308–309.

4

Distorted Support: Pathologies of Weather Warfare

James R. Fleming

A mutually supportive relationship has long existed between science and the military in the United States. In the nineteenth century scientists often "hitched a ride" with army and navy exploring expeditions; field sciences utilized military scope, organization, and discipline to collect their data; and military institutions forged their identities in part around scientific agendas, leadership, and training. In military circles, the prestige of science grew, linked as it was to scientists' power over nature, both actual and perceived. Scientists as well gained access to recognition and political power through channels such as the new, but largely honorific, National Academy of Sciences and new federal agencies identified with scientific services.

This relationship expanded and became more permanent in the twentieth century. The National Research Council coordinated scientific research during America's brief involvement in World War I; and during World War II the Office of Scientific Research and Development (OSRD) and the Manhattan Project demonstrated to the government and the military the absolute importance of promoting and supporting scientific research, development, testing, and evaluation of new weapons systems. Beyond the crush of particular conflicts, science in the Cold War era and beyond has become a prominent and permanent component of the American military.[1]

Given such continuing support, integration, and identification between science and the military—in perspectives, values, budgets, scale, and personnel—what does the term "distortion" mean? Hasn't this large-scale relationship, the pursuit of power in modern states, grown up over the centuries gradually yet inexorably to become just the way things are? How can we say that science or the military have somehow been perverted, misshapen, or twisted out of their "proper" or "natural" positions or relationships?

My interests involve exploring pathological aspects of otherwise seemingly mutual and beneficial relationships.[2] A physician investigating a skin lesion or breast tumor for cancer is unlikely to test a patient for heart palpitations or digestive disorders; nevertheless, a suspicious lump or mole in an otherwise "healthy" person can be the defining event in a person's life (or death).

The three cases presented here: meteorology in the postbellum, World War I, and Cold War eras are examples of a relationship forged over the course of centuries between science and its state patrons. In this relationship it is safe to say that scientists seek support from the state and access to political *power*, while the state (especially the military) seeks *powers* over nature as promised (and sometimes delivered) by scientists. As geophysicists pursue the knowledge of the earth they have long sought, they willingly focus their investigations on those areas of knowledge where technology and military interests have made resources readily available. The examples presented here represent protochapters in a longer work that explores military-science relationships using examples drawn from the geosciences—examples that focus on enthusiasms or pathologies, either hidden or beyond the pale of "normal" support and interaction. These relationships were forged—far beyond simple patronage—with roots much deeper than we might have imagined.

STORMS, STRIKES, AND SURVEILLANCE

Following the Civil War, the federal government established a massive and well-funded national weather service with direct links to the military. Although weather observations had been made in America since colonial times, the first national service providing daily weather reports and forecasts was established from 1870 to 1891 in the Signal Service, in effect extending the career of its leader, Gen. Albert James Myer, and ensuring the postwar continuation of the agency.[3]

In addition to their more mundane meteorological duties, the weather service, linked to Washington by military and commercial telegraph networks, served as a national surveillance force, reporting to the government on a variety of threats to the domestic order such as striking railroad workers, Indian uprisings on the frontier, locust outbreaks, and natural hazards to commerce and agriculture.[4]

It can be argued that, at the time, the weather mission was just as important to the Signal Service as military support was to the development of meteorology. Nevertheless, as the Signal Service gathered meteorological data to provide "telegrams and reports for the benefit of commerce and agriculture," its military commanders were engaged in a much larger project to gather intelligence on all possible threats to domestic tranquillity—presiding over an information network that embodied the physical, intellectual, and cultural resources (and fallibilities) of the society that constructed it. For example, President Rutherford B. Hayes had telegraphic reports every three hours from around the country from the Signal Service during the railway strikes of 1877, and Indian fighters had special telegraph lines built by the Signal Service into Apache and Sioux territories. It was not a completely smooth relationship. According to William A. Glassford, a contemporary chronicler of the Signal Service who knew Myer personally, "To the *legitimate* duties of military signaling he added the *utterly foreign* concerns of a meteorologist."[5]

As I have recently argued elsewhere, the postbellum U.S. Army Signal Office, with its emphasis on surveillance and intelligence gathering regarding storms, strikes, Indian uprisings, and other threats to domestic order, was a precursor of today's Office of Homeland Security.[6] General Myer, who focused on surveillance during his tenure as chief signal officer from 1870 to 1880, was succeeded by Gen. William Hazen and Gen. Aldophus Greely, who were not nearly as aggressive in pursuing this agenda. Still, they commanded an elite force of college-educated weather officers, exempt from basic training and required to pass a stringent entrance examination involving French and German translations, mathematics, meteorology, and a request to "enumerate the titles of any considerable scientific publications of which you are the author, in whole or in part."[7] This, and the investigation of the role of science in the army by the Allison Commission in 1885, opened the Signal Service to widespread suspicion by the regular army.[8]

By 1891 the weather service was moved to the Department of Agriculture. Only a year later, however, the Signal Office, now a purely military outfit, advised the National Guard on the use of signaling devices—heliographs, lanterns, and telegraph lines—in the Homestead riots in Pennsylvania at the Carnegie Steel and Iron Company. During the Chicago railroad riots of 1894, the Signal Office installed a "very complete" system of visual, telephone, and telegraph communications between department commanders and subordinates. The lessons Myer learned during the rail strike of 1877 were not lost on his successors. Although officially "forgotten" today, they have been reconstituted as homeland security.

METEOROLOGY IN WORLD WAR I

Military meteorology reentered the picture in World War I when the Signal Corps, headed by Brig. Gen. George O. Squier, recruited the noted physicist Robert A. Millikan to provide meteorological support for the war effort as head of the new "Army Meteorological and Aerological Service."[9] With support from the Weather Bureau (now part of Agriculture), and in concert with their European colleagues, meteorologists supported new fields of study on aviation meteorology, "battlefield climatology," and local weather conditions conducive to launching or surviving poison gas attacks.

The impact of aviation on World War I, although huge, was late in arriving in America. As historians Charles Bates and John Fuller have pointed out, "When the Germans swept across the plains of Flanders in August, 1914, nearly reaching Paris, America's political and military leaders doubted that aviation could play a meaningful role in warfare."[10] Six months later, the United States had only 2 percent of the world's pilots and a handful of military aircraft on the runway. In the final months of the war, however, several hundred officers and enlisted men received their meteorological and aerological training. Like their nineteenth-century Signal Corps predecessors, they constituted a technical elite and were exempt from front-line infantry service.[11]

American meteorologists specialized in upper-air observations and wind trajectories. In the newly minted field of aeronomy, data collection from balloons, airships, and airplanes supported reconnaissance and mail flights. Visibility, which played such a crucial role in the naval battle of Jutland, was now a primary concern for aviation. Some speculated that upper-air winds could guide balloons over enemy territory to drop propaganda leaflets.[12] Yet the primary role of aeronomy was in calculating the effects of upper-air winds on the trajectories of long-range artillery shells. A contemporary training manual noted that wind was not to be considered a "capricious element." Although it increases with height and "becomes tangent to the isobars at an altitude about five hundred to one thousand meters," it is, except during the passage of depressions, "approximately the same over a considerable area (and notably one as great as the front of an army)."[13] Nevertheless, while "windage and elevation" entered the artillerists' vocabulary, forward spotters and trial and error served to guide the big guns more than esoteric and complex meteorological computations.

Information about turbulence, stability, local winds, and other characteristics of the "boundary layer" near the ground was systematized for the use to pilots and artillerists, but was also critical to the infantry in their preparations to launch or defend against poison gas attacks.[14] In this sense, the fields of micrometeorology and microclimatology, largely agricultural in their nineteenth-century origins, had effectively been militarized.

World War I was dominated by the crushing realities of trench warfare and the controlling influence of generals "mud" and "winter." This, however, did not dampen the enthusiasm of promoters like Alexander McAdie, head of the new Aviation Weather Advisory Service of the U.S. Navy, who crowed, "Who commands the air, commands all! . . . Henceforth the ships of the sky shall play the leading role, and the nation holding the mastery of the air will have in its palm the power to make or mar. . . . The strategies of warfare will be entirely different. Individuals will count for less; machines and *weather* will determine the victory."[15]

COLD WAR CLOUD SEEDING

In the Cold War era, techniques of modifying the weather and possibly the climate developed by Irving Langmuir and his associates at the General Electric Corporation led to crash programs in military cloud seeding in search of a weather weapon that could be used surreptitiously to release the violence of the atmosphere against an enemy, tame the winds in the service of an all-weather air force, or, on a larger scale, perhaps disrupt (or improve) the agricultural economy of nations and alter the global climate for strategic purposes.[16] If researchers were right, a small amount of "nucleating" agent, such as dry ice, silver iodide, or even water, could cause a "chain reaction" in clouds that would release as much energy as an atomic bomb, but without radioactive fallout; moreover, it would be unidirectional in that clouds seeded over Europe would be carried by the westerly winds over the Soviet Union.

Planners generated strategic scenarios that included hindering the enemy's military campaigns by causing heavy rains or snows to fall along lines of troop movement and on vital airfields, or using controlled precipitation as a delivery system for biological or radiological agents. Tactical possibilities included dissipating cloud decks to enable visual bombing attacks on targets, opening airfields closed by low clouds or fog, and relieving aircraft icing conditions. In military circles cloud seeding was seen as the trigger that could release the violence of the atmosphere against an enemy or tame the winds in the service of an all-weather air force.

The technology seemed of such great potential, especially to military aviation, that Vannevar Bush, a friend of Langmuir and former head of the Office of Scientific Research and Development in World War II, brought the issue to the attention of Secretary of Defense George C. Marshall and Gen. Omar Bradley, chairman of the Joint Chiefs of Staff, who immediately convened a committee headed by meteorologist Sverre Petterssen to serve as a buffer between the defense establishment and the scientific community while research proceeded on secret weather weapons.[17]

The classic Cold War pronouncement on weather control belongs to Gen. George C. Kenney, commander of the Strategic Air Command: "The nation which first learns to plot the paths of air masses accurately and learns to control the time and place of precipitation will dominate the globe."[18] But he was not alone. The distinguished aviator-engineer Rear Adm. Luis De Florez, who developed synthetic training devices for Navy fliers during World War II, had a similar opinion: "With control of the weather the operations and economy of an enemy could be disrupted. . . . [Such control] in a cold war would provide a powerful and subtle weapon to injure agricultural production, hinder commerce and slow down industry." De Florez urged the government to "start now to make control of weather equal in scope to the Manhattan District Project which produced the first A-bomb."[19]

Pursuing this theme, Harold Orville, President Dwight D. Eisenhower's weather adviser, published an influential 1954 article in *Collier's* that included scenarios for using weather as a weapon of warfare. Planes would drop hundreds of balloons containing seeding crystals into the jet stream. Downstream, when fuses on the balloons exploded, the crystals would fall into the clouds, initiating rain and miring enemy operations. The Army Ordnance Department was investigating another technique: loading silver iodide and carbon dioxide into .50 caliber tracer bullets that pilots could fire into clouds. A more insidious technique would strike at the enemy's food supply by seeding clouds to rob them of moisture before they reached enemy agricultural areas.[20]

Although in Orville's assessment, total weather mastery would be possible only after four decades of intensive research, the spin-offs from this work, when combined with the maturation of electronic computers, would provide a *completely accurate* system of weather forecasting, perhaps within a decade: "I think it entirely probable that, in 10 years, your daily weather forecast will read something like this: 'Freezing rain, starting at 10:46 A.M., ending at 2:32 P.M.' or 'Heavy snowfall, seven inches, starting today at 1:43 A.M., continuing throughout day until 7:37 P.M.'"[21] This sort of accuracy of prediction, even without weather control, would have major consequences for military operations. Although speculative and wildly optimistic, ruminations such as these from an official source and threats that the Soviets were aggressively pursuing weather control helped fuel a weather race with the Russians and the rapid expansion of meteorological research in all areas, but especially in weather modification.

Ironically, in 1953, at the very same time Langmuir was himself involved in making highly dubious claims for the efficacy of weather and even climate modification, he presented a seminar at GE titled "Pathological Science" or "the science of things that aren't so."[22] Utilizing his own criteria for pathology, acceptance of Langmuir's claims for cloud seeding apparently involved overreliance on the credentials of a Nobel laureate, rather than proof.

CLOUD SEEDING IN SOUTHEAST ASIA

Weather warfare took a macropathological turn between 1967 and 1972 in the jungles over North and South Vietnam, Laos, and Cambodia.[23] Under Operation POPEYE, the Air Weather Service conducted secret cloud seeding operations to reduce trafficability along portions of the Ho Chi Minh Trail.

Flying out of Udorn Air Base, Thailand, without the knowledge of the Thai government or almost anyone else, but with the full and enthusiastic support of President Lyndon B. Johnson,[24] the Air Weather Service flew more than twenty-six hundred cloud seeding sorties and expended forty-seven thousand silver iodide flares over a period of approximately five years at an annual cost of approximately $3.6 million.

In March 1971 nationally syndicated columnist Jack Anderson broke the story about air force rainmakers in Southeast Asia in the *Washington Post,* a story confirmed several months later in the *Pentagon Papers*.[25] A year later Secretary of Defense Melvin Laird stonewalled the Senate Foreign Relations Committee when he claimed there was no cloud seeding going on over *North Vietnam,* never mentioning that Operation POPEYE was still functioning elsewhere in the region.[26] Nevertheless by 1973 the Senate had adopted a resolution (S. 71) "to prohibit and prevent, at any place, any environmental or geophysical modification activity as a weapon of war," and in 1974 Sen. Claiborne Pell forced the declassification of the transcript of a top-secret briefing by the Defense Department on the topic.

Operation POPEYE, made public as it was at the end of the Nixon era, was called the Watergate of weather warfare.[27] While some argued that environmental weapons were more "humane" than nuclear weapons, others suggested that inducing rainfall to reduce trafficability was preferable to dropping napalm; as one wag put it, "make mud, not war." Philip Handler, president of the National Academy of Sciences, represented the mainstream of scientific opinion, however, when he observed: "It is grotesquely immoral that scientific understanding and technological capabilities developed for human welfare to protect the public health, enhance agricultural productivity, and minimize the natural violence of large storms should be so distorted as to become weapons of war."[28]

The Soviet Union, realizing the weakness of the U.S. position on cloud seeding in Vietnam and taking full advantage of the Watergate crisis, seized the diplomatic initiative and caused considerable embarrassment to the Ford administration by bringing the issue of weather modification as a weapon of war to the attention of the United Nations. The UN Convention on the Prohibition of Military or Any Other Hostile Use of Environmental Modification Techniques, entered into force in 1978, ironically, when the Lao People's Democratic Republic, where the American military had used weather modification technology in war only six years earlier, became the twentieth nation to ratify it.[29]

CONCLUSION

As these examples demonstrate, meteorology in America was undoubtedly shaped by its political, administrative, and military history. More than simple cases of military support or patronage for science, these are stories of the interpenetration of values and perspectives among meteorologists and military officers in distinct eras: following the Civil War, during World War I, in the early Cold War, in Vietnam, and in current military circles.

Has the interaction of science and the military simply fulfilled a Faustian bargain struck in the early modern era, if not before? Weapons systems of the past century—increasingly based on science—have also been increasingly lethal (especially to civilians), increasingly toxic, and increasingly pathological. Physics, chemistry, and biology have weaponized the atom, molecule, virus, and bacterium, while the geosciences have militarized the global environment in the air, under the seas, and in outer space. Can science in the service of the military ever take a permanent, nonlethal turn and become an agent of peace, perhaps in medicine, passive restraints, or smart weapons of greater precision? This is not likely in a world still filled with nuclear weapons, rogue states, ethnic hatreds, terrorists, ad hoc and clumsy alliances, and at least one self-righteous superpower.

What lessons can be drawn from the historical examples presented here?

1. Telegraphy, a technology that enabled surveillance of enemy troops in the Civil War, allowed surveillance of the entire nation, especially of undesirable groups, in peacetime. A national scientific service—a national weather service—provided the infrastructure, in effect "homeland security" in the 1870s.
2. Aviation, the wave of the future identified with military power, required the enhanced services of meteorology and aeronomy. Knowledge of lift, lob, and loft were all needed for planes, shells, and poison gas, all of which rode the air currents. A desire to alter the weather, especially fogs, for the benefit of pilots began as early as World War I, while climatology and local meteorology became militarized.

3a. Cold warriors presumed that clouds, storms, and even the climate, like any other natural phenomenon, could be controlled and weaponized. They further presumed that the Soviets were probably ahead of the United States on this.

3b. During the Johnson and Nixon administrations—the era of generation gaps and credibility gaps—it was assumed that the surreptitious use of environmental warfare was acceptable, as long as you didn't get caught. One observer noted that the lesson of the Vietnam experience was not that rainmaking is an inefficient means for slowing logistical movement in jungle trails, but "that one can conduct covert operations using a new technology in a democracy without the knowledge of the people."[30] The dominant opinion, however, was that seeding clouds—like using Agent Orange or the Rome Plow, setting fire to the jungles, or bombing the dikes over North Vietnam—was but one of many sordid genocidal and ecocidal techniques used in Vietnam.[31]

Although these examples are widely separated in time, they represent a much larger class of pathological interactions between modern science and the military. In an age of terrorism and global battlefields, they may actually point to a trend in scientific practice rather than a distortion of it.

NOTES

1. Barton C. Hacker, "Military Patronage and the Geophysical Sciences in America: An Introduction," in James Rodger Fleming, ed., special theme issue on geophysics and the military, *Historical Studies in the Physical and Biological Sciences* 30, no. 2 (2000): 1–5.

2. Irving Langmuir first articulated his definition of pathological science in a colloquium in 1953; for a more recent version, see Irving Langmuir, "Pathological Science," transcribed and edited by Robert N. Hall, *Physics Today* (October 1989): 36–48.

3. Donald R. Whitnah, *A History of the United States Weather Bureau* (Urbana: University of Illinois Press, 1961); Joseph M. Hawes, "The Signal Corps and Its Weather Service, 1870–1890," *Military Affairs* 30 (1966): 68–76; James Rodger Fleming, *Meteorology in America, 1800–1870* (Baltimore, MD: Johns Hopkins University Press, 2000).

4. James Rodger Fleming, "Storms, Strikes, and Surveillance: The U.S. Army Signal Office, 1861–1891," *Historical Studies in the Physical and Biological Sciences* 30, no. 2 (2000): 315–32.

5. W. A. Glassford, "The Signal Corps," in *The Army of the United States*, ed. Theophilus F. Rodenbough and William L. Haskin (New York: Maynard, Merrill, 1896), 136–49.

6. James Rodger Fleming, "Telegraphing the Weather: Military Meteorology, Strategy, and 'Homeland Security' on the American Frontier in the 1870s," in *Instrumental in War: Science, Research, and Instruments between Knowledge and the World*, ed. Steven A. Walton (Leiden: Brill, 2005), 153–78.

7. Army Signal Office, Special Examination, Questions for Competitive Examination before the United States Civil Service Commission, in *Annual Report of the Chief Signal Officer of the Army to the Secretary of War for the Year 1885*, 48th Cong., 1st Sess., H. Exec. Doc., vol. 8, no. 1, 1885, pp. 77–79.

8. "Testimony before the [Allison] commission to consider the present organization of the signal service . . ." U.S. Sen., Misc. doc. 82, 49th Cong., 1st Sess. See also A. Hunter Dupree, *Science in the Federal Government: A History of Policies and Activities* (Cambridge, MA: Harvard University Press, 1957), 190, 215, et seq.

9. Charles C. Bates and John F. Fuller, *America's Weather Warriors, 1814–1985* (College Station: Texas A&M University Press, 1986), 18. Two brief official accounts are George O. Squier, "Meteorological Service of the Army," *Monthly Weather Review*

47, no. 1 (1919): 84; and U.S. Weather Bureau, "The Signal Corps Meteorological Service," *Monthly Weather Review* 47, no. 12 (1919): 870–71. See also Philip M. Flammer, "Meteorology in the United States Army, 1917–1935," MA thesis, George Washington University, 1958.

10. Bates and Fuller, *Weather Warriors*, 16.

11. Oliver L. Fassig, "A Signal Corps School of Meteorology," *Monthly Weather Review* 46, no. 12 (1918): 560–62.

12. Bates and Fuller, *Weather Warriors*, 21.

13. *Notes on the Meteorologic Elements Affecting Artillery*. Translation of "Note sur les elements météorologiques interessant l'artillerie," Army War College, U.S. War Department, Doc. no. 688 (Washington, DC: Government Printing Office, 1917), 12–13.

14. John F. Fuller, *Thor's Legions* (Boston: American Meteorological Society, 1990), 12; Bates and Fuller, *Weather Warriors*, 23.

15. Alexander McAdie, "War Weather Vignettes," in *Alexander McAdie: Scientist and Writer*, comp. Mary R. B. McAdie (Charlottesville, VA: M. R. B. McAdie, 1949), 296, 261; Alexander McAdie, "Making the Weather," 325–26.

16. This section is based on James R. Fleming, "Fixing the Weather and Climate: Military and Civilian Schemes for Cloud Seeding and Climate Engineering," in *The Technological Fix*, ed. Lisa Rosner, Hagley Center Studies in the History of Business and Technology (New York: Routledge, 2004), 175–200.

17. *Weathering the Storm: Sverre Petterssen, the D-Day Forecast and the Rise of Modern Meteorology*, ed. James Rodger Fleming (Boston: American Meteorological Society, 2001).

18. Frank L. Kluckhohn, "$28,000,000 Urged to Support M.I.T."*New York Times*, 15 June 1947, 46, 1.

19. Arthur Krock, "An Inexpensive Start at Controlling the Weather," *New York Times*, 23 March 1961.

20. Howard T. Orville, "Weather Made to Order?" *Collier's* (28 May 1954), 25–26.

21. Orville, "Weather Made to Order?" 26; original emphasis preserved.

22. Langmuir, "Pathological Science," 36–48; Robert G. Fleagle, "Second Opinions on 'Pathological Science,'" *Physics Today* (March 1990): 110.

23. U.S. Cong., Sen., Comm. on Foreign Relations, Subcomm. on Oceans and International Environment, *Weather Modification: Hearings*, 93rd Cong., 2d Sess., 1974, 87ff. The operational phase began on 20 March 1967, and was conducted each year during the monsoon season (March–November) until 5 July 1972. The project was also known as Intermediary Compatriot, and, by AWS, as Motorpool. See John F. Fuller, *Air Weather Service Support to the United States Army: Tet and the Decade After*, AWS Historical Study No. 8 (Scott AFB, IL: Military Airlift Command, 1979), 30–32.

24. As early as 1957, then-Senator Johnson had speculated on controlling the earth's weather from space for military purposes; see Lowell Ponte, "Weather Warfare Forecast: Partly Cloudy—U.N. Treaty Would Permit 'Peaceful' Environmental Research by Military," *Los Angeles Times*, 29 January 1976; reprinted in U.S. Cong., Sen. Comm. on Foreign Relations, Subcomm. on Oceans and International Environment, *Prohibiting Hostile Use of Environmental Modification Techniques: Hearing*, 94th Cong, 2d Sess., 1976.

25. Jack Anderson, *Washington Post*, 18 March 1971. Memorandum from Deputy Secretary of Defense to the Hon. Nicholas deB. Katzenbach, Under Secretary of State, Subject: "Military Action Program for SE Asia," 21 February 1967, cited in Department of Defense, *United States-Vietnam Relations, 1945–1967: Study Prepared by the Department of Defense*, book 5, vol. 2, *U.S. Ground Strategy and Force Deployments: 1965–1967* (Washington, DC: Government Printing Office, 1971), 50–51. See also Bates and Fuller, *America's Weather Warriors*, 229–32.

26. Testimony of Melvin R. Laird before the Senate Foreign Relations Committee, April 18, 1972; cited in *Weather Modification Hearings*, 109–110; see also Daniel S. Greenberg, "Vietnam Rainmaking: A Chronicle of DoD's Snowjob," *Science and Government Report* 2 (1 May 1972): 1–4.

27. Cristine Russell, "The Weather as a Secret Weapon: From Vietnam to Geneva," *Washington Star*, 23 August 1975; reprinted in U.S. Cong., Sen. Comm. on Foreign Relations, Subcomm. on Oceans and International Environment, *Prohibiting Military Weather Modification: Hearings on S.R. 281*, 92d Cong., 2d Sess., 1972, 47.

28. Philip Handler to Claiborne Pell, July 25, 1972, printed in *Prohibiting Military Weather Modification*, 153.

29. United Nations, *Multilateral Treaties Deposited with the Secretary-General: Status as of 31 December 1982* (New York: Author, 1983), 667. The text of the UN Convention (A/RES/31/72) is reprinted as appendix C in Congressional Research Service, *Weather Modification: Programs, Problems, Policy, and Potential*, 95th Cong., 2nd Sess., 1978, 510–13.

30. Gordon J. F. MacDonald, statement in U.S. Cong., House Comm. on International Relations, Subcomm. on International Organizations, *Prohibition of Weather Modification as a Weapon of War: Hearings on H.R. 28*, 94th Cong., 1st Sess., 1975, 5.

31. For a detailed account of American environmental war efforts and their effects in Indochina, see Stockholm International Peace Research Institute, *Ecological Consequences of the Second Indochina War* (Stockholm: Almqvist & Wiksell, 1976). For background, see Ruth Russell, "The Nature of Military Impacts on the Environment," in Sierra Club, *Air, Water, Earth, Fire* (San Francisco: Sierra Club Special Publication, 1974, 1–14.

5

Dreadnoughts and Elections: America's Response to Naval Revolution

Walter Earl Pittman

Between 1900 and World War I, naval technology underwent changes so revolutionary that they rivaled the earlier transitions to steam-driven steel warships by the world's navies. These changes included development of entirely new ship types such as the dreadnought battleships, battle cruisers, and submarines, as well as the continued evolution of destroyers and cruisers into ship types that remained standard for decades. Gunnery improved in accuracy, range, and destructiveness by more than an order of magnitude.

The introduction of steam turbines allowed larger ships to be driven at higher speeds with lower fuel consumption. Oil replaced coal as the primary fuel. Radio linked far-flung fleet units by 1914. The evolution of torpedoes, mines, shaped charges, smokeless powder, super-hardened armor plate, and ship designs was accelerated. Organizational structures were rationalized as the major powers moved toward centralized command and control systems integrated with the political leadership and centered on a general staff system supported by war planning, intelligence, and research and development functionalities. The most significant of these changes were encompassed in the designs of the dreadnought battleships. These changes took place in an era of heated great power rivalry. None of the great powers dared to be left behind, and the consequent naval race became a major factor leading to war.

In the United States, the naval buildup was driven by fear of German and Japanese intentions toward America's interests in Latin America, the Open Door in China, and its newly acquired colonial possessions. Suspicion of Germany grew out of the incidents in Manila Bay in 1898 that convinced Adm. George Dewey and others of Teutonic aggressiveness. This distrust permeated the navy and strongly influenced Theodore Roosevelt.

Apprehensions hardened as Adm. Alfred von Tirpitz's huge naval building program posed an ever-increasing threat to American interests. In the Venezuelan Crisis of 1902–1903, German intervention confirmed popular suspicions. In this case, the Germans were forced to abandon their aggressive designs in the Caribbean by a superior American naval force.[1]

Japan emerged from the Russo-Japanese War as the dominant power and the leading aspirant to Far Eastern hegemony. Only World War II would end its renewed encroachments into China. The United States, as the traditional supporter of Chinese integrity and the possessor of extremely vulnerable colonial real estate in the Philippines, quickly found itself in lonely opposition to Japanese expansion. Japanese-American relations were exacerbated almost to the breaking point in 1906 and 1907 by a series of incidents in the Pacific states. Japanese pride suffered from the exclusion of Japanese students from San Francisco's public schools, and anger grew with riots and boycotts of Japanese merchants and laborers in all the western slope states. President Roosevelt, a young man new to the job, walked a thin line, caught between West Coast demands for total Japanese exclusion and Japanese insistence upon fair treatment for their countrymen. Judiciously blending threats to use the "big stick" with heavy doses of calming salve, he avoided a major confrontation, for the moment.[2]

America's enlarged international commitments and the growing naval power of its potential foes created major problems for the new president. The president and much of the public accepted Alfred Thayer Mahan's concepts of sea power and recognized the need for a powerful naval force to protect American interests and possessions. But opposition was significant. The Democrats, particularly in the South, had traditionally opposed naval growth since Jefferson's day. They were joined by midwestern congressmen who preferred cheaper commerce raiders, at first cruisers and later submarines, for Mahan's expensive battle fleet.

Congress habitually authorized ships one year and appropriated funds for construction at the next session, which resulted in an average building time of five years.[3] The navy lacked men as well as ships. Even ammunition was scarce, with no emergency stockpiles, and congressional appropriations were too small to allow enough ammunition for adequate training.[4] However stingily Congress provided the essentials of national defense, it was often wildly extravagant in maintaining useless naval stations. It also insisted upon the retention of an archaic and cumbersome bureaucracy whose primary virtue was that it allowed for political meddling in the procurement of equipment and supplies.[5]

Furthermore, Congress had a tendency to ignore the recommendations of the professional military leaders and instead to write military appropriations the way domestic legislation was written. This unwieldy system of compromise and logrolling led to naval forces that were too weak to uphold national commitments and that were unbalanced as weapon systems. Congress paid little attention to the real-world conditions in which the fleet would have to operate and which should have determined the optimum strength of the navy. The result was that despite spending $400 million more than Germany, the U.S. Navy was no more powerful by 1914.[6] The civilian leaders in the government were of little help. None of the secretaries of the navy (John D. Long, Dwight Moody, George von Lengerke Meyer, Josephus Daniels) had the kind of scientific or engineering background that might have enabled them to understand the technical issues they had to decide. None had any extensive knowledge of foreign affairs or much interest in naval reform or growth. Long and Daniels, in fact, were hostile to changes. Adm. Bradley A. Fiske, aide for operations under presidents William H. Taft and Woodrow Wilson, concluded that the intellect of cabinet members was decidedly inferior to the senior naval officers, engineers, and scientists that he otherwise dealt with.[7] The exceptions were Theodore Roosevelt, who took a deep personal interest and a hands-on approach to military technology, and his spokesman in Congress, Richmond P. Hobson, a former naval architect. Roosevelt determined many of the important technical decisions (army small arms, dreadnought designs, aircraft). He was also the first president to submerge in a submarine.[8]

Roosevelt was effective in increasing the size of the navy before 1905. Aided by the confrontation with Germany over Venezuela, he persuaded Congress to authorize five new battleships in 1903. The navy's power was enhanced by the traditional efficiency of the U.S. Navy and the fact that the United States normally built the most powerful ships of a given class in the world. It was a navy designed for offensive operations in the Mahan tradition.[9]

Roosevelt's success in attaining naval increases came in the face of increasing opposition as the public's memory of the Spanish-American War faded. Pacifism and faith in the efficacy of international agreements, such as the Hague Conferences of 1899 and 1907, and in international law and government grew in the public and Congress. The movement culminated in Hamilton H. Holt and William H. Taft's "League to Enforce Peace" in 1915, from which Woodrow Wilson derived his League of Nations concept.[10] The peace movement was based on denial of threats to U.S. interests and offered an alternative to those opposed to increasing armaments on ideological or economic grounds. It became strong enough that even the navalist leaders like Roosevelt had to give at least lip service to the movement. In fact, the president was the moving spirit in convening the Second Hague Conference in 1907. Privately, he was "utterly disgusted with the nonsense chattered by the extreme advocates of peace."[11] But, ever the realist, Roosevelt understood what he had to do to assure his naval program.

Between 1906 and 1908, the growing peace movement, naval technological innovation, and heated international tensions resulted in a political collision in America. It was precipitated by the secret building, in less than a year, of the British super battleship, HMS *Dreadnought*. It was impelled by concerns over gunnery accuracy in the U.S. Navy and the Royal Navy of Great Britain. While the public had been gratified by the quick, efficient destruction of the Spanish fleets at Manila Bay and Santiago, the navy had been stunned by postbattle studies showing that although fought at near point-blank range in bright light in calm seas (the Spanish ships were anchored in Manila Bay), only 1.3 to 2.3 percent of the shells fired struck their targets. A young naval officer, William S. Sims, inspired by the gunnery reforms of Sir Percy Scott of the Royal Navy, introduced new systems of loading, pointing, and firing and provided or inspired numerous technical innovations. Continuous aiming techniques were introduced that required more precise and rapidly adjustable train and elevation mechanisms and the retraining of every gun crew in the navy. Rapid loading techniques were introduced, which required new type of ammunition hoists and loading mechanisms. Training was changed to emphasize speed as well as accuracy of firing. Telescopic sights and coincidence range finders were introduced. Automatic data control systems to transmit targeting data from fire control centers to gun turrets were developed as well as mechanical analog computers to solve the gunnery equations governing the flight of shells between two moving ships. Sims won Roosevelt's support by approaching him directly outside of channels and his reforms quickly reached throughout the fleet. By 1917, U.S. Navy ships were achieving 40 to 60 percent hits in practice at ranges up to thirty thousand yards, rather than the 1 to 3 percent hits at three thousand yards or less of the 1898 war.[12]

Other problems were evident. Adjustment of gunfire required spotting of the shell's impact point. This is inherently difficult when there are a number of guns of varying calibers firing at a target. The *Connecticut* class of 1902 carried four twelve-inch guns, twelve seven-inch guns, and twenty-eight smaller guns.[13] Salvo firing, from a ship whose entire battery consists of the same size guns, makes spotting easier by creating only a single large shell splash to spot. There were also seen to be advantages in placing fire control centers as high up as possible in a ship so that spotters could see farther over the horizon. Accuracy in range estimation was also a function of size: The larger the separation between two base sight points, the more accurate the range estimates. So there were inherent advantages in larger ships. Some theorists believed that future naval battles would be decided by massive fire concentrations of big guns firing from battleships maneuvering at maximum range, which they could maintain because of their higher speeds. Attaining these higher speeds required new power systems because the triple-expansion reciprocating steam engines that drove the world's battleships could not maintain high speed without suffering mechanical failure. The answer was, according to the reformers, the steam turbine, which was already in use in smaller vessels. A transition to oil from coal for fuel was also urged. It provided greater range and flexibility, and after 1912 Great Britain and the United States committed themselves to oil.[14]

Naval architecture had been gradually moving toward the dreadnought design before 1905. In 1903 Roosevelt urged the building of an all big gun battleship, but the Bureau of Naval Construction raised objections. The 1905-class ships, *South Carolina* and *Michigan*, were sixteen-thousand-ton battleships armed with eight twelve-inch guns. They were, however, too slow and too limited in range to be classed as dreadnoughts. In any event, they were not completed until 1910. America's first dreadnought was *Delaware*, twenty-thousand-ton displacement, 21.5-knot speed, armed with ten twelve-inch guns, which was not completed until 1910.[15]

It was the only battleship that the president asked for in 1905 and was probably the only one he could have gotten. The peace movement was at the peak of its strength, basking in popular approval, and the public, enjoying a period of prosperity, was uninterested in foreign alarms. But the outside world could not be ignored, and growing tensions with Japan and instability in the European balance of power kept Roosevelt's attention focused on the defense needs of the nation. The Battle of Tsushima changed Roosevelt's concerns about American naval power to something close to panic.

At Tsushima on 27 May 1905, a smaller Japanese fleet intercepted the Russian Baltic Fleet in the decisive battle of the Russo-Japanese War. Utilizing his fleet's seven-knot speed advantage, Adm. Heihachiro Togo stood off at maximum range, and repeatedly pounded the Russian battle line with concentrated main-battery fire until he destroyed the Russian force, sinking twenty warships, including four new battleships, with only minor losses to his force. And, although Mahan tried to argue otherwise, at great cost to his own reputation, the concepts that had lead to the dreadnought design were proven unmistakably.[16] As a war scare gripped the nation, President Roosevelt found himself in a crisis with Japan, a navy that was suddenly obsolescent, and a Congress unwilling to change things.

Roosevelt turned to the naval question with his characteristic energy and determination. In his fight, he gained an ally and a new floor leader in Congress: Alabama congressman Richmond P. Hobson, a naval hero of the Spanish-American War who had been warning of the Japanese threat for years.[17] An extremely effective speaker, Hobson reached numerous influential Americans through the Chautauqua circuits, where he was one of the most popular and highest-paid attractions. Those out of reach of his voice could read his words on the pages of William Randolph Hearst's newspapers and magazines.[18]

Using his war-won fame as a springboard into politics, Hobson entered Congress in 1906 with the primary purpose of increasing America's naval strength in what he saw as an increasingly dangerous international environment. In his four-term congressional career, he never wavered from this goal and soon emerged as the navy's leading spokesman on Capitol Hill. In fact, Hobson, a southern Democrat, became the Republican Roosevelt's spokesman on naval matters in his first term in office, an almost unprecedented occurrence.[19]

Roosevelt began a carefully orchestrated campaign aimed at the 60th Congress, hoping for major naval increases in 1908. Specifically, he asked Congress to abandon the annual political wheeling and dealing that traditionally characterized naval appropriation acts. Instead, Roosevelt asked for the adoption of a long-range plan of naval building. He wanted Congress to endorse the proposals of the navy's General Board that called for creation of a fleet of forty-eight battleships, which would be the largest fleet in the world. The board's plan required the building of four battleships a year until completion. Roosevelt then asked for two things: a congressional commitment to a long-range building program and the authorization for four new battleships in 1908.[20]

The president's first step in his campaign was to order the Atlantic Fleet to circumnavigate the world. Although also aimed at impressing the world—particularly Japan—with American naval might, Roosevelt's primary motive was to shift domestic and congressional opinion in favor of naval expansion. The cruise of the Great White Fleet succeeded in monopolizing headline space during the entire congressional naval battle of 1908.[21] The second step was made by Congressman Hobson, who mailed circular letters to the mayors and editors of local papers in each of the towns where he had lectured. Copies of the letters, which were propaganda for naval increases, were also sent to each member of Congress. Admiral Dewey also entered the fray with a series of public statements urging naval increases and warning that war might come if the United States did not arm.[22]

The fireworks began in the House Naval Affairs Committee, which first considered the administration's request for four new battleships and equivalent auxiliaries. The opposition was fierce because the president was trying to get Congress to tie its annual naval appropriations to a long-range program of matching German and Japanese naval increases. After bitter debate, the committee rejected the president's program and recommended the construction of only two battleships. The number of submarines, of very limited range and effectiveness—they were labeled "coast defense submarines"—was doubled. Hobson fought hard for the president's proposal and stubbornly refused to compromise. He prepared a minority report pleading for four battleships, but was the only one to sign it. He carried his fight to the floor of Congress, but despite what was described as "one of the most impassioned speeches ever heard in the House," his minority report was overwhelmingly rejected. Repeated efforts by Hobson to amend the appropriations bill on the floor of the House to allow for either four or even three new battleships also failed.[23]

"Yesterday the House seemed to possess an infinite capacity to go wrong," Roosevelt sadly wrote to Hobson. But in the Senate, the navalists fared no better. Administration forces led by Sen. Albert J. Beveridge won a moral victory over the conservatives but were unable to get more than the two battleships approved by the House. The Senate vote was also decisive.[24]

The navalist forces failed despite extreme efforts. In the midst of the congressional struggle, Hobson embarked upon a nationwide speaking tour. Roosevelt used extraordinary pressure on individual congressmen, issued a dramatic public appeal for support, and even threatened to veto pork barrel legislation in his effort to obtain four battleships. But Congress was not moved. Ironically, polls of newspaper editors seemed to indicate that the navalists had better luck reaching the public than they did reaching Congress.[25]

There were many factors involved in the defeat of Roosevelt's naval program. The strongest congressional opposition came from the Midwest, while public opposition was greatest in the East among intellectual, business, and pacifist groups. The *Houston Post* and the *St. Louis Globe-Democrat* pointed out that the defeat of the large-navy bill was in part due to congressional fear that enlarged naval appropriations would force a curtailment in pork barrel spending. Roosevelt's threat to veto pork barrel legislation in an election year was simply not credible. Other congressmen opposed the increased appropriations because of the high costs involved. Many of these congressmen were also pacifists who opposed extensive American involvement in international affairs.[26]

The year 1908 was also an election year and the conservative Republicans were both weary of the "big stick" and rebellious. As a lame-duck president, Roosevelt was unable to crack the party whip hard enough to ensure passage of his measures. The Democrats, led by John Sharp Williams, expressed the traditional Democratic aversion to a large navy and also seemed to distrust Roosevelt's motives for wanting more warships.[27]

The basic cause of the defeat of the Naval Bill of 1908 was the failure of Hobson, Roosevelt, and other navalists to convince Congress that a real emergency existed in regard to Japan. Had Congress been convinced that war was imminent, it would have given the president the necessary arms; any other course of action could have been construed as unpatriotic and hence political suicide.

There is no doubt that Roosevelt seriously feared that war was imminent. The president confided his apprehension to Hobson, who promptly blurted it out at the Democratic Platform Committee meeting in Denver. A national uproar ensued. Privately furious, Roosevelt denied the statement and broke off personal relations with Hobson.[28] But they both had failed to transfer their sense of impending doom to Congress.

The optimists among the navalists later pointed out that the entire battle had not been lost. By demanding four battleships, they received funds for two and defeated the one-a-year replacement policy Congress favored. The *Army and Navy Register* believed the navy had obtained a "reasonable increase" and pointed out that the authorization for two battleships created no precedent in the face of congressional failure to adopt a definite long-range program.[29] They were whistling in the dark, for despite the continuation of major efforts by navalist leaders through the years, Congress stubbornly supported the two-battleships-a-year pattern until the sudden emergency of World War I found the United States still unprepared.

NOTES

1. Seward W. Livermore, "Theodore Roosevelt, the American Navy and the Venezuela Crisis of 1902–1903," *American Historical Review* 53 (1958): 422; Holger H. Holwig, *Politics of Frustration: The United States in German Naval Planning, 1889–1941* (Boston: Little, Brown, 1976), 67–85; Harold Sprout and Margaret Sprout, *The Rise of American Naval Power 1776–1918* (Princeton, NJ: Princeton University Press, 1939), 289–91.

2. Outten J. Clinard, *Japan's Influence on American Naval Power 1897–1917* (Berkeley and Los Angeles: University of California Press, 1947), 47–51; *Jackson (MS) Clarion Ledger*, 26 May 1907, 30 May 1907; Raymond L. Buell, "The Development of the Anti-Japanese Agitation in the United States," *Political Science Quarterly*, 37 (1922): 605–22.

3. U. S. Cong., *Navy Yearbook, 1909*, Sen. Doc. 5, 61st Cong., 2nd Sess., 1909, 647–50.

4. *Army and Navy Register* 39 (26 May 1906): 6.

5. C. W. Stockton to Stephen B. Luce, 21 March 1907; Luce to William Sims, 28 August 1908; to Theodore Roosevelt, 14 July 1908 (Stephen B. Luce Papers, Library of Congress).

6. President of the General Board of the Navy to the Secretary of the Navy, 3 October 1907, 20 December 1907 (General Board Records, Navy Historical Center); C. W. Stockton to Stephen B. Luce, 21 March 1907 (Luce Papers).

7. Bradley A. Fiske, *From Midshipman to Rear-Admiral* (New York: Century, 1919), 528, 535.

8. Matthew M. Oyos, "Theodore Roosevelt and the Implements of War," *Journal of Military History* 60 (1996): 631–56.

9. "Strength of the Naval Powers," *Scientific American* 93 (8 July 1905): 26; *Army and Navy Register* 39 (24 March 1906): 6.

10. Warren F. Kuehl, *Seeking World Order: The United States and International Organizations to 1920* (Nashville, TN: Vanderbilt University Press, 1969), 75–297.

11. Theodore Roosevelt to Hermann Speck von Sternberg, 16 July 1907, in *The Letters of Theodore Roosevelt*, ed. Elting E. Morison (Cambridge, MA: Harvard University Press, 1952), 5:366.

12. Elting E. Morison, *Admiral Sims and the Modern American Navy* (Boston: Houghton Mifflin, 1942); *Army and Navy Register* 34 (26 June 1906): 6; *Army and Navy Register* 34 (16 June 1906): 6; S. S. McClure, "Admiral Sims," *McClure's Maga-*

zine 47 (Nov. 1922): 26–31; Peter Padfield, *Aim Straight* (London: Hodder & Stoughton, 1966), 59–222.

13. *Navy Yearbook, 1909*, 650–55.

14. Morison, *Admiral Sims*, 97–136. Kenneth Wimmel, *Theodore Roosevelt and the Great White Fleet* (Washington: Brassey's, 1998), 149–91. Robert K. Massie, *Dreadnought: Britain, Germany and the Coming of the Great War* (New York: Random House, 1991), 373–497; Nicholas Lambert, *Sir John Fisher's Naval Revolution* (Columbia: University of South Carolina Press, 1999).

15. *Navy Yearbook, 1909*, 658–59.

16. W. S. Sims, "Tactical Qualities of the *Dreadnought* Class," *Scientific American Supplement* 42, 26008–12.

17. *New York Herald*, 29 April 1900; *Washington Times Magazine*, 27 October 1907; *Washington Times Magazine*, 3 November 1907; Richmond P. Hobson, "If War Should Come," *Cosmopolitan* 55 (June 1908): 34.

18. M. L. Williamson to R. K. Johnson, 2 May 1935 (Richmond P. Hobson Papers, Library of Congress); *North Vernon (IN) Sun*, 6 October 1911; *Lonville (NY) Herald*, 17 February 1916.

19. *Jackson (MS) Clarion Ledger*, 13 March 1908; *U.S. Congressional Record*, 60th Cong., 1st Sess., 1908, 40, part 1, 450; *U.S. Congressional Record*, 63rd Cong., 2nd Sess., 1914, 51, part 1, 87, 89–91; part 7, 7148, 7151–52.

20. *U. S. Congressional Record*, 60th Cong., 1st Sess., 40, part 1, 81.

21. Theodore Roosevelt to Anna Cabot Mills Lodge, 20 September 1907, in *Letters of Theodore Roosevelt*, 5:858–59; Roosevelt to William Bayard Hale, 3 December 1908, in *Letters of Theodore Roosevelt*, 6:1408.

22. Circular letter in the Hobson Papers; *New York Times*, 15 March 1908.

23. *U. S. Congressional Record*, 60th Cong., 1st Sess., 1908, 43, Part 5, 4574–82, 4607, 4611, 4612–4777, 4807.

24. Theodore Roosevelt to Richmond P. Hobson, 16 April 1908 (Theodore Roosevelt Papers, Library of Congress); *U. S. Congressional Record*. 60th Cong., 1st Sess., 1908, 42, Part 6, 5291.

25. *Spokane (WA) Daily Chronicle*, 27 April 1908; *Los Angeles Express*, 8 May 1908, 10 May 1908; *Jackson (MS) Clarion Ledger*, 15 April 1908; Sprout and Sprout, *Rise of American Naval Power*, 266.

26. "Two Battleships or Four?" *Literary Digest* 36 (2 May 1908): 631–32. Sprout and Sprout, *Rise of American Naval Power*, 267–68.

27. *U. S. Congressional Record*, 60th Cong., 1st Sess., 1908, 42, part 6, 4789–94.

28. *Boston American*, 10 July 1908; Press release, 9 July 1908 (Roosevelt Papers); Theodore Roosevelt to Richmond P. Hobson, 9 July 1908; Hobson to Roosevelt, 1 August 1908, 9 August 1908 (Hobson Papers).

29. "Congress and the Services," *Army and Navy Register* 44 (6 June 1908): 6; ibid. (9 May 1908): 7.

6

Testing Democracy: First World War IQ Testing, from Measuring the Military to Selecting the Student Body

Mark Soderstrom

In 1937 Malcolm MacLean, dean of the University of Minnesota General College and a founder of the university guidance program, wrote a letter to his friend Malcolm G. Little. In the course of the letter, MacLean mentioned his friend's son (also named Malcolm), who came to Minnesota for educational guidance. MacLean discussed the young man's prognosis, based on the results of a whole battery of tests he had taken that took two pages to list. A few were academic tests, but personality and intelligence tests made up the bulk of the battery, each one designed to quantify a different specific aspect of the prospective student's persona and rank it against those of his peers.[1] Yet at the beginning of the twentieth century, the idea that intelligence was measurable at all was widely met with derision in education and industry.[2] How did an entire regime of tests become enshrined in the process of student life at a land-grant university a mere twenty years later?

I focus here on events at the University of Minnesota in the Twin Cities, but variations of the same story played out across the country between the world wars, as universities sought to manage ballooning enrollments. World War I set in motion two chains of events that involved the university; the two together would combine to create the academic testing culture that would come to permeate the campus. The admission of students through the Student Army Training Corps, or SATC, overwhelmed administrators with a sudden influx of students that the university was unprepared to handle. At the same time, the army was trying out new technology to sort its soldiers, the Army Alpha Intelligence Test. The legitimacy that such large-scale use gave intelligence testing, combined with academic administrators' own imperative to winnow postwar student populations, led to the development of a testing regime on campus that sought to restrict admissions in an "impartial" and politically acceptable manner.

The initial effect of World War I on campuses was diminished enrollments, resulting in financial difficulties. Many campuses throughout the country, and many departments at the University of Minnesota in particular, cut back faculty.[3] Despite these problems, administrators and faculty were exceptionally pro-war—per capita, more so than the U.S. population at large—and sought ways to serve the war effort.[4] When the army proposed the SATC in the summer of 1918, academic administrators were thrilled, both by the very real relief it posed for their own bottom line, and by the prospect of being of service to the nation. "There is some small satisfaction in feeling that . . . we may be doing a little something to assist in prosecuting the war," wrote Cephas Allin, University of Minnesota professor of political science. "The man in the trenches, the miner and the farmer seem to be so much more valuable members of society than we could possibly be."[5]

As details of the program became better known, however, initial academic enthusiasm began to wane. The SATC was set up to be a holding pool of personnel from which the army could draw at will, and who meanwhile would be receiving higher education, training either to be officers or technical specialists.[6] To maximize that pool, the SATC rewrote admissions standards to universities on a national scale, allowing any able male high school graduate (or man with an equivalent education) to matriculate.[7] University of Minnesota administrators were not pleased by the overriding of their own selective entrance requirements, and were disturbed that the Minneapolis superintendent of schools had chosen to publicize the new standard in Minneapolis newspapers.[8]

The SATC imposed many adjustments on the university system. Schools that had cut back faculty due to previous low enrollments were given only one month's notice to prepare for a student body larger than they had ever had before: six thousand new students from the SATC alone. Professors wrote new war-related courses and textbooks nearly overnight. Students were housed in hastily commandeered barracks off campus.[9] Military discipline was imposed on the student body and the campus: Their military officers assigned SATC students KP, drill, and guard duty. The schedules for such duties were made without regard to class times or responsibilities, causing absenteeism and a lack of time for study that proved very frustrating to the academic faculty and university administration.[10] So too did SATC students' habit of compensating for the rigors of their training by sleeping "as though under enchantment straight through their classes, in the library and over their books at night."[11]

Such behavior merely reinforced faculty and administrators' notion that open enrollment allowed unfit students to waste the educational opportunity extended by the university.[12] To add insult to injury, normal academic hierarchy was inverted as student guards were told by their military officers to demand identification from their administrators and faculty, treating them as civilians at a military encampment at the entrances and exits of their own university buildings.[13] The University of Minnesota set a committee to study the effects of the SATC on school functions, and the subsequent report documented "the tremendously vital problem of the interference by military authorities with the educational activities of the University and of the members of the S.A.T.C."[14] By the end of the war, the same Cephas Allin wrote that "the SATC was an educational affliction."[15]

The actual SATC program did not last very long: It was announced in August 1918, and put into place in October. Scarcely a month later came the Armistice, followed by the program's decommissioning at the end of December.[16] Odious problems for the university lingered, though. Schools still struggled with massive numbers of students now left without funding halfway through the year. Nor could the precedent of open enrollment easily be withdrawn. The next year, 1919, the university hired extra faculty in expectation that postwar enrollment would increase. Administrators estimated a thousand incoming freshmen and thought that was erring on the generous side.[17] They failed to recognize the sea change the SATC had wrought. The American Council on Education had promoted the SATC heavily in the late summer of 1918, using the slogan "it's patriotic to go to college."[18] For the first time in U.S. history, college was linked broadly to citizenship and patriotism, and portrayed as accessible.

The result was that instead of the thousand freshmen expected by the university in 1919, twice that number sought entrance.[19] At the same time, as James Gray writes, returning servicemen resumed their studies and veterans made use of free tuition to give higher education a try. Total enrollment in 1919 rose to eight thousand students, almost a quarter again as much as the peak prewar enrollment of sixty-three hundred. Faculty and administration alike were overwhelmed, and the university faced serious financial difficulty.[20] One dean likened the influx to barbarians at the gates of Rome.[21] Administrators truly believed that such open access to university education was wasted on people who would be better off working, rather than permitted to entertain false hopes of educational attainment that would be inevitably dashed. Administrators' dismay at the demands placed on their institution was exacerbated by a sense of futility at wasting limited state funds and university resources on students fated to fail.[22]

At the same time, high school administrators waged an effective public relations battle to ensure that the previous admission standards would not return. They wrote impassioned letters to the editors of local newspapers and to the university calling any reinstatement of prewar admissions requirements old-fashioned and undemocratic.[23] University administrators, for their part, were keenly aware that they depended on public funding and goodwill. In order to preserve that good feeling in the public realm, administrators sought a new screening method that would distinguish merit and educability on some objective scale, thereby addressing the democratic concerns of opportunity while still achieving their own goal of limiting access. University deans found the answer in another wartime technology that was developed with University of Minnesota resources: the Army Intelligence Test.

As Daniel Kevles has shown, the validity of intelligence as a quantifiable entity was mostly treated with disdain as late as the eve of World War I. Only psychologists and others in related fields had much faith in intelligence testing. But the war changed all that, for two important reasons. When the army administered intelligence tests to almost two million inductees, the very scale of the activity actually legitimated it in the public eye. At the same time, the sheer mass of data that had been collected pacified more academic doubters. By the end of the war, businesses and schools alike were clamoring to use intelligence testing to sort their own personnel and students.[24]

Minnesota had a head start in this process, largely because of two men: Dean John Black Johnston of the College of Science, Literature, and the Arts (and former chair of the Psychology Department), and Robert Yerkes, the man Johnston had selected to replace him as chair of Psychology and the man who had convinced the army to administer the Alpha Tests in the first place. Yerkes was already a nationally known figure in his field, having taught at Harvard, served as president of the American Psychological Association (APA) in 1917, and published significant research both in intelligence testing and animal behavior. Working through the APA and the National Research Council with colleagues like C. C. Bingham and Lewis Terman, he had persuaded the army early in the war to test all recruits in order to best sort their suitability for service. By 1917 the military was confronted with an unprecedented mass of recruits, and the APA's testing program offered the possibility of quickly and scientifically determining which branch, level, and field of the military would best suit each new soldier.[25] Although the actual utility of the tests for the army was not very high in the wartime period, it did produce some notable results. During the course of the war, the officer corps multiplied by a factor of twenty-one, from nine thousand to two hundred thousand. Of those officers, two-thirds had come from training camps that used the army intelligence tests to rank recruits.[26]

The first set of college students given the Army Alpha Test as a matter of routine nationally were in fact the SATC participants, as part of the military effort to grade the students on a number of factors in order to determine their potential as officers.[27] Therefore, educators already had some knowledge of the procedures involved in large-scale testing. The validity of subjecting all students to standardized rating was established by association with the military, and therefore became validated as an objective technique of college selection. Educators began to expand the scope of their testing to younger students, and by 1921, four million public school students in the United States had been given intelligence examinations.[28]

A look at the Army Alpha Test reveals what determiners were used to measure intelligence and fitness for officer training. On the first page, Test One, recruits were given ten seconds to complete commands that corresponded with visual cues. For example:

> Attention! Look at 4. When I say "go" make a figure 1 in the space which is in the circle but not in the triangle or square, and also make a figure 2 which is in the triangle and circle but not in the square. Go.
>
> Attention! Look at 6. When I say "Go" put in the second circle the right answer to the question: "how many months are there in a year?" In the third circle do nothing, but in the fourth circle put any number that is a wrong answer to the question you have just answered correctly. Go."[29]

The second page of the exam, Test Two, consisted of word problems, many of which were oriented toward military situations: "A certain division contains 2,000 artillery, 15,000 infantry, and 1,000 cavalry. If each branch is expanded proportionately until there are in all 19,800 men, how many will be added to the artillery?" Other questions focused on digging trenches or the speed of trains, sounding very much like word problems from any elementary school.[30]

Test Three was multiple choice, with questions focused on what was assumed to be common knowledge: "Gold is more suitable than iron for making money because . . ." Possible answers were: "Gold is pretty," "Iron rusts easily," or "Gold is scarcer and more valuable." (Spartans would have disagreed with all of the above, preferring iron for their money.) Another question reached toward commonsense economics: "Why do some men who could afford to own a house live in a rented one?" Possible answers were: "They don't have to pay taxes," "They don't have to buy a rented house," and "They can make more by investing the money the house would cost."[31]

Test Four was a vocabulary test, asking examinees to choose whether two words were synonyms or antonyms; choices included such pairs as "desuetude-disuse" and "lugubrious-maudlin." Test Five asked examinees to unscramble a sentence and choose whether the sentence was true or false. One example: "a general not major a and rank same the of are." Test Six asked the examinees to predict the next number from a given series.[32]

Test Seven was an analogy test which would be familiar to generations of SAT takers: "dog—bark::cat—" with possible answers of "chair," "mew," "fire," and "house."[33] The final test, Test Eight, was another series of multiple-choice questions. This test included a few odd questions, such as "How many legs does a Korean have?" but primarily this part of the test seems to have been designed to determine a certain level of knowledge that was not so much a matter of innate intelligence as cultural background. It asked where Yale was, and gave four college towns as choices. It asked questions about agriculture, humanities, automobiles, and golf.[34] In this test, examiners sought to find themselves in the recruits and students, rewarding a certain minimum level of exposure to the kind of rural bourgeois life that David F. Noble has identified as the background of most Progressive engineers, scientists, and academics.[35]

For the military, one advantage of the test overall would be that the test modeled cooperative intelligence. As Joel H. Spring demonstrates, "[I]ntelligence [as defined by the Alpha and Beta Tests] was an indication of the ability to function in modern corporate forms of activity, such as the army, the factory, and the school. . . . The model was the beehive, where everyone performed a task suited to his capabilities."[36]

While the testers claimed that the test measured innate intelligence or capacity, as Stephen Jay Gould and others have established, the design, implementation, and interpretation of the test biased the results against immigrants, children of immigrants, working-class people, and African Americans.[37] However, this aspect of the tool would not necessarily have been a negative in the eyes of academic administrators seeking to retain an exclusionary prewar status quo. In fact, administrators (and test designers such as Yerkes) were blind to this aspect of the exam, seeing only that it produced results that confirmed the meritocratic order as expected.

For his part, Dean Johnston had been an early proponent of testing, and as early as 1915 he became interested in using tests to ascertain fitness for college.[38] A psychologist specializing in neurological anatomy, Johnston was committed to a view that the university was a system whereby worthy individuals could be trained to best serve society as leaders, and he resented any diversion of state educational resources for those who would waste them through failure or selfish choices. He considered unmerited aristocracy to be as much a drain on the system as academic failure, and looked to the scientific determination of intelligence as a way to confer education upon those most suited to make use of it. There was a political aspect to his desire to be more selective, as well: He argued that students who had failed or dropped out were "centers of criticism and disaffection toward the University" at a time when the university was seeking to expand its state influence and resources.[39] He was interested in an admissions system that was publicly acknowledged to be fair, in order to limit the public relations damage that numbers of failed students might wreak. Testing was his answer.

Johnston administered intelligence tests to the entering class of the university medical school in 1915 as a data-gathering endeavor.[40] The outbreak of war gave him the opportunity he needed to refine his agenda. In a 1917 talk entitled "The Duty of the University during the War," Johnston suggested that the university emulate the selective draft, stressing the ideal of the draft as the will of the people over the lives of individuals. Using a draft-like approach, the university would select only those students "who are best fitted to perform the service for which the University exists."[41]

For Johnston, testing was an ideal way to discern the worthy, and his alliance with Yerkes, who was already working on intelligence tests for the army when Johnston recommended him for Psychology chair, was an ideal way to move the University of Minnesota into the testing vanguard. By 1917 Johnston was hard at work on a Minnesota version of the test, thanks to Yerkes, who supplied his home department with copies of the army test in exchange for the data collected.[42] The university was therefore the first educational institution in the nation to have access to the test for its own purposes. Even though at the time possession or duplication of the Army Alpha Test was illegal and punishable by a hefty fine and jail time, the only penalty Johnston suffered was a tart reminder from the Surgeon General's Office in March 1918.[43] In fact, just the reverse, for by the time the war was over and other schools and private businesses were clamoring to develop testing procedures of their own, Johnston and Yerkes had assured Minnesota a substantial head start. This lead meant that Minnesota would have a national (and international) role in developing the new technology of mass testing, eventually culminating in the development of the Minnesota Multiphasic Personality Inventory, or MMPI.[44]

In 1917 Johnston was already using the Army Alpha Test on all university freshmen, as well as selected public school students all the way down to the fifth grade.[45] Well before any other school had even seen the army test blanks, Johnston and his Minnesota colleagues were already developing the staff and procedures for large-scale testing, amassing their own store of data, and planning their own refinements to the test. In 1922 Johnston published a follow-up study on the University of Minnesota seniors whom he had tested as freshmen in 1917, correlating their freshman test results with their subsequent college achievement.[46] Johnston and his colleagues concluded that the army test, while it did correlate with college success to some degree, was an insufficiently precise instrument for testing college entrants, geared as it was for a much broader population.

The large-scale implementation of high-stakes intelligence tests soon called for refinements. Administrators asked for more, and more specific, tests to track other criteria or differentiate ranks of intelligence more finely. New tests were developed, including the Minnesota College Ability Test in 1921.[47] Johnston experimented with numerous supplementary admission criteria, including high school grades and qualitative statements from students and their high school principals, to flesh out early test results.[48] Within a few years, though, he dropped the qualitative measures and settled on a combination of intelligence test scores and high school rank to provide his measure of college aptitude.[49] Johnston developed yet another index based on a comparison of rank and intelligence, thus determining "moral" or "motivational" factors. He intended the index to identify lazy students whose high school records did not measure up to the potential indicated by their intelligence tests.[50]

High school principals and superintendents, who had argued so publicly against restrictive college entrance requirements during and after the war, were more persuaded that testing was a fair metric for college suitability. Indeed, testing enhanced high schools' role in the educational process, since they became the sole source of information for Johnston's admission guidelines. Principals administered the tests to their students and were thus brought into the system as full participants. As the testing regime became more established, and the efficacy and ethics of state surveillance of the individual became more accepted, the amount and intimacy of the data collected on students at all stages in their school lives increased. Parents and students themselves became participants in the tracking and exclusion process. In the summer of 1925 Johnston mailed letters to high school seniors' parents, with advice based on their children's test performance, and noted that as a result fewer students from lower ranks entered the university that fall.[51]

Not all of the initial proponents of testing were pleased with the potentially exclusionary direction the movement was taking. Melvin E. Haggerty, dean of the College of Education at the University of Minnesota and a former Yerkes student,[52] had originally been one of the forces behind testing students. By 1928, though, he had publicly split from Johnston and the other Minnesota testers, criticizing Johnston for his vision of university education as "limited to the two percent of the population that want to enter the learned professions."[53] Where Johnston saw the good of society as paramount over the good of the individual, Haggerty believed that education should focus on the needs of the individual and the development of individual access to opportunity, from which the collective good would necessarily follow. To Haggerty, testing was not so much a tool to exclude unfit students as a tool to identify how best to help students learn. He saw intelligence as dynamic and multiple rather than fixed and singular, and envisioned a much broader role for university education than just a training ground for leaders.[54]

Both the army and the university were social institutions adjusting to the imperatives of modern mass society. For both, testing assisted in determining the individual's best place inside their respective organizations. To the appropriately trained expert, testing objectively revealed an individual's true nature. The expert was thus in a position to tell the individual more about him- or herself than the individual could possibly know. The development of student guidance resolved the Progressive tension between society and the individual, as the expert was able to liberate the individual by identifying the individual's exact niche in the social or organizational hierarchy.

As early as 1922, Lotus D. Coffman, the new president of the University of Minnesota, instituted a program of guidance and orientation to complement the intelligence testing that was already being performed.[55] Johnston advocated guidance even in high schools, to help with the task of sorting students who would benefit from college from those who would not, and providing vocational advice to those who were not college material.[56] Minnesota psychologists developed tests to assay personality traits, such as vocational ability, mental hygiene, social adjustment, and artistic or creative aptitude.

In 1924 Harold S. Diehl, MD, who worked in the medical school's Department of Preventive Medicine, began testing all freshmen to look for mental pathology. His test asked questions ranging from "Are you bothered much by blushing?" and "Do you lack the usual interest in the opposite sex?" to "Do you know of anybody who is trying to do you harm?" Students whose answers indicated possible "neurotic tendencies" were interviewed personally, and some were referred to counseling. Overall Diehl estimated that 10 percent of all university students had "quite real problems," and in addition a "fairly large number [of students beyond that 10 percent] should receive some aid."[57] In 1932 the University Testing Bureau was formed to coordinate the growing body of intelligence and aptitude tests administered to students.[58] By 1934 the bureau director advocated testing high school students yearly for intelligence and aptitude, accumulating this data in a permanent record, offering guidance based on the tests, and forwarding the record on to business, industry, or college, depending on where the student went after school.[59]

And so we find ourselves again in 1937, and reading Malcolm MacLean's letter about his friend's son. After administering his set of tests, MacLean was able to tell his friend more about his son than the son ever could have said about himself. Here are just a few examples from the two-page list that MacLean detailed.

> We gave him our battery of social adjustment scales and found him, as we expected, in the 90th percentile in adequate social behavior, and in the 70th in social preferences—in other words, finely adjusted socially. These attitude tests showed, also, a considerably better than average morale—the good old guts to you—and courage; a slight sense of inferiority, which is natural and, I think, a good thing, since it keeps him humble and not either cocky or foolhardy. . . . He showed excellent attitudes toward his home and family on both the Rundquist-Sletto Scale and the Bell Adjustment Inventory. . . . In his attitude towards law he was about midway. . . . Economically we found him quite a radical, being in the

> 87th percentile on the radical side. Insofar as these tests are adequate, we find him also emotionally adjusted and with a good clear sensible attitude towards his own health. In maturity of interests for his age 20, we find him 8 percentile points above the median, which is natural and all to the good. On our very interesting new test for masculinity and femininity, we find him in the 81st percentile on the masculine side—hence, no panty-waist, no pansy, and yet with enough of the feminine so that he won't be a hard-boiled bird.
>
> On the Strong Vocational Interest Test, it was extremely interesting to note that he had no absolute peak identities with any one group. He stood out clear and strong in the teaching and personnel fields, ranking a B in one and a B+ in the other. . . . [H]e showed also B identity with Y. M. C. A., physical director, and Boy's Secretary patterns. He also had B ratings in science and math, and very low B in the business field and medicine. None of the other thirty or more occupations showed any color in him at all.
>
> In consequence, the general field of personnel is pretty clearly indicated.[60]

World War I was a turning point for American academia. The promise of the Morrill Act was realized. Due to the SATC, the entire basis by which public universities operated, including whether their mandate gave them a right to select students or gave them the responsibility to accept all comers, was called into question. And yet the curse also provided the cure: World War I testing technology proved readily adaptable to the needs of Progressive administrators wanting to exert authority over their institutions, and even provided educators and administrators with a way to reconcile their democratic ideals with their managerial impulses. At the end, the situation had become a boon for those academics, who enjoyed the cachet of impartiality and promised new service in individual adjustment to social needs through the technology of selection and guidance.

NOTES

1. Malcolm S. MacLean to Malcolm G. Little, 22 September 1937 (Box 1, "Little, Malcolm G." folder, Malcolm McLean Collection, University Archives, University of Minnesota, Minneapolis).
2. Daniel Kevles, "Testing the Army's Intelligence: Psychologists and the Military in World War I," *Journal of American History* 55 (1968): 566.
3. James Gray, *The University of Minnesota: 1851–1951* (Minneapolis: University of Minnesota Press, 1951), 249–50.
4. Carol S. Gruber, *Mars and Minerva: World War I and the Uses of the Higher Learning in America* (Baton Rouge: Louisiana State University Press, 1975), 81–82.
5. Cephas D. Allin to R. M. Alden, 13 September 1918 (Minnesota Department of Political Science Correspondence, University Archive, University of Minnesota, Minneapolis), as cited in Gruber, *Mars and Minerva*, 232.
6. Gruber, *Mars and Minerva*, 216–17.

7. Gruber, *Mars and Minerva*, 217. See also "S. A. T. C. Telegram No. 4," to President Burton, 10 September 1918 (Box 12, "Subject File: Student Army Training Corps, 1917–1919" folder, Minnesota University Science Literature and the Arts Collection, University Archives, University of Minnesota, Minneapolis [hereafter referred to as the MNSLA Collection]).

8. President Marion L. Burton, "Bulletin to the Deans," 10 September 1918 (Box 12, "Subject File: Student Army Training Corps, 1917–1919" folder, MNSLA Collection).

9. James Gray, *Open Wide the Door: The Story of the University of Minnesota* (New York: G. P. Putnam's Sons, 1958), 122.

10. Frederick J. Wulling to Dean J. B. Johnston, 23 December 1918 (Box 12, "Subject File: Student Army Training Corps, 1917–1919" folder, MNSLA Collection).

11. Gray, *Open Wide the Door*, 122–23.

12. Gray, *Open Wide the Door*, 122–23.

13. Gruber, *Mars and Minerva*, 234–37.

14. President Marion L. Burton, "Bulletin to the Deans," 8 November 1918 (Box 12, "Subject File: Student Army Training Corps, 1917–1919" folder, MNSLA Collection).

15. C. D. Allin to R. M. Alden, 26 June 1919 (Minnesota Department of Political Science Correspondence, University Archives, University of Minnesota, Minneapolis), as cited in Gruber, *Mars and Minerva*, 237.

16. University of Minnesota, *The President's Report 1918–1919* (Minneapolis: University of Minnesota Press, 1919), 9–14.

17. John Black Johnston to Superintendent B. B. Jackson, 1 November 1919 (Box 10, "Subject File: Entrance Requirements 1919–1927" folder, MNSLA Collection).

18. Gruber, *Mars and Minerva*, 228.

19. Johnston to Jackson, Minneapolis, 1 November 1919.

20. Gray, *Open Wide the Door*, 124.

21. J. B. Johnston, "Predicting College Achievement as a Basis for Educational Guidance," *Educational Record Supplement* 2 (July 1926): 36.

22. Johnston, "Predicting College Achievement," 33–47.

23. See, e.g., Principal R. T. Hargreaves to Dean J. B. Johnston, 3 November 1919, and Superintendent John Munroe to Dean J. B. Johnston, 4 November 1919 (both from Box 10, "Subject File: Entrance Requirements 1919–1927" folder, MNSLA Collection).

24. Kevles, "Testing the Army's Intelligence," 565–81.

25. Kevles, "Testing the Army's Intelligence," 565–81.

26. Kevles, "Testing the Army's Intelligence," 573.

27. Clarence S. Yoakum to J. B. Johnston, Washington, DC, 31 January 1919 (Box 6, "Committee: Mental Tests, 1917–1920" folder, MNSLA Collection).

28. Melvin E. Haggerty, "Recent Developments in Measuring Human Capacities," *Journal of Educational Research* 3 (1921): 242.

29. Stephen Jay Gould, *The Mismeasure of Man* (New York: W. W. Norton, 1996), 230.

30. Group Examination Alpha Blank, 2.

31. Group Examination Alpha Blank, 3.

32. Group Examination Alpha Blank, 4–6.
33. Group Examination Alpha Blank, 7.
34. Group Examination Alpha Blank, 8.
35. David F. Noble, *America by Design: Science, Technology, and the Rise of Corporate Capitalism* (New York: Alfred A. Knopf, 1977; reprint, New York: Oxford University Press, 1979), 50–51.
36. Joel H. Spring, "Psychologists and the War: The Meaning of Intelligence in the Alpha and Beta Tests," *History of Education Quarterly* 12 (1972): 13.
37. Gould, *Mismeasure of Man*, 222–63.
38. Johnston, "Predicting College Achievement," 33.
39. John Black Johnston to Commission on Institutions of Higher Education of the North Central Association, 20 March 1918 (Box 10, "Subject File: North Central Association 1918" folder, MNSLA Collection).
40. Donald G. Paterson, "Educational Guidance at the University of Minnesota," 21 February 1924 (Box 5, "Committee: Educational Guidance 1922–1924" folder, MNSLA Collection).
41. John Black Johnston, "The Duty of the University during the War," chapel talk, 4 December 1917 (John Black Johnston Papers, University Archives, University of Minnesota, Minneapolis, 3–5).
42. Johnston, "Predicting College Achievement," 33. For the data exchange arrangement, see John Black Johnston to Maj. Robert Yerkes, 8 March 1918, and John Black Johnston to Professor J. M. Thomas, 18 December 1917 (both from Box 6, "Committee: Mental Tests 1917–1920," MNSLA Collection).
43. Robert M. Yerkes to John Black Johnston, 5 March 1918 (Box 6, "Committee: Mental Tests 1917–1920," MNSLA Collection).
44. Gray, *University of Minnesota*, 358.
45. Paterson, "Educational Guidance," and Johnston to Yerkes, 8 March 1918.
46. Paterson, "Educational Guidance."
47. John B. Johnston, *The Liberal College in Changing Society* (New York: Century, 1930), 117–18.
48. Johnston, "To the Commission on Institutions of Higher Education."
49. Gray, *Open Wide the Door*, 100.
50. John Black Johnston, "Predicting College Achievement as a Basis for Educational Guidance," *Educational Record Supplement* 2 (1926): 40–41.
51. John Black Johnston to Dr. Adam Leroy Jones, 16 October 1925, and John Black Johnston to Dr. Edgar L. Yeager, 16 October 1925 (Box 3, Folder 8, MNSLA Collection).
52. Gray, *University of Minnesota*, 222, 472.
53. M. E. Haggerty, "Address to the MEA," 1 October 1928 (Box 1, "Speeches 1920–1921" folder, Melvin Haggerty Papers, University Archives, University of Minnesota, Minneapolis), 6.
54. Haggerty, "Address to the MEA," 1–20.
55. Paterson, "Educational Guidance."
56. J. B. Johnston, "The Determination of Fitness for College as an Element in Vocational Guidance," (paper presented at the annual meeting of the Vocational Guidance Association, Detroit, MI, 21 February 1931), as sent to Dean Helen Bragdon, University of Rochester, on 14 January 1931 (Box 4, Folder 18, MNSLA Collection).

57. Harold S. Diehl, "Preliminary Report of Subcommittee on Mental Hygiene," 15 December 1924 (Box 5, "Committee: Educational Research, 1924–1935 "folder, MNSLA Collection).

58. Gray, *University of Minnesota*, 353.

59. E. G. Williamson to John Black Johnston, 13 February 1934 (Box 13, "Subject File: Ass'n of Minn. Colleges. Committee on Application for Admission Blank 1933–1935" folder, MNSLA Collection).

60. MacLean to Little, 22 September 1937.

7

Dr. Albert B. Sabin, the U.S. Army, and the Conquest of Epidemic Disease

John M. Morra

Dr. Albert B. Sabin's fame rests principally on the development of a live, attenuated oral polio vaccine. According to the World Health Organization, the Sabin vaccine has been in large measure responsible for the expected global eradication of paralytic poliomyelitis by 2008.[1] The great medical achievement of eradication, of course, was the work of many individuals, most notably Jonas Salk, with whom Sabin is inevitably linked. Sabin, by most measures, was the single major figure in the polio drama. He helped clarify the natural history of poliovirus infection as entering primarily through the mouth and multiplying in the alimentary tract. He found three avirulent strains that could be administered orally, preventing paralysis, and breaking the chain of transmission of the viruses. He developed the mass immunization strategy that has been effective in conquering polio. For this alone he would be remembered as a great scientist.[2]

Yet Sabin was also an indefatigable researcher in many areas of virology, and made other important contributions to medical science. Many of these contributions were made when Sabin was an officer in the United States Medical Corps during World War II, and afterward in his continuous role as adviser to the army on the Armed Forces Epidemiological Board until 1973. Sabin's army connection spurred a lifelong interest in the field of tropical medicine and proved exceptionally creative.[3] Moreover, Sabin's army experiences carried over into his polio work in several ways and proved to be one factor in a most significant part of the polio drama: Russian medical researchers used Sabin's poliovirus strains to vaccinate eighty million people from 1958 to1960. The American oral vaccine's effectiveness was shown by resolute Soviet scientists.[4]

Sabin became an adviser to the Army Epidemiological Board (later called the AFEB) in 1939, and continued in that role until 1973. He joined the Army Medical Corps officially in early 1943 after a year of processing delays. Sabin wanted to enlist after Pearl Harbor, but the Medical Corps insisted that all research be done in Army facilities. Several research projects were underway at Sabin's Children's Hospital Medical Research Foundation laboratory in Cincinnati that he did not want to leave uncompleted. Early in 1943, the army created a number of special commissions that allowed prominent researchers to work in both kinds of facilities.[5] Sabin, like Dr. Tom Rivers, the widely acknowledged father of American virology, and his friends and fellow polio researchers Frederick Robbins, Joel Warren, and William Hammon, was in the Medical Corps while simultaneously advising it from the outside. Thus, Sabin's army experiences coexisted with and overlapped his great polio breakthrough of the 1950s and 1960s.

By the time he entered the Medical Corps at age thirty-six, Sabin was already an important figure in American science with a long list of achievements. As a student at New York University Medical School, he developed a rapid diagnostic test that identified the type of bacteria that brought a patient with pneumonia to the hospital. It was called the Sabin pneumococcal test, and was a question on the New York State Medical Exam, which, ironically, Sabin was required to answer in applying for his medical license. From 1934 to 1939, he worked at the Rockefeller Institute in New York under Tom Rivers, who also directed polio research for the National Foundation for Infantile Paralysis (NFIP), the March of Dimes. Sabin isolated the B-virus that usually kills individuals bitten by infected monkeys. He did his early polio work elucidating how the disease really worked. This contradicted the esteemed Dr. Simon Flexner, who held that polio entered through the nose and spread first to the brain. As mentioned earlier, Sabin proved the poliovirus entered primarily through the mouth and multiplied in the intestinal tract. In 1948 he also diagnosed the dangerous parasite toxoplasmosis and developed the Sabin/Feldman blue dye test to identify it. This led to the discovery that toxoplasma infection of pregnant women was a major cause of congenital brain disease in newborn infants.

Always interested in public health as well as medical research, Sabin noticed that children who underwent tonsillectomies when polio was present in the summer plague time were much more likely to develop paralytic polio of the bulbar kind that required life support in a respirator.[6] Thus the practice of avoiding such operations in the summer became common. He also warned of the link between very heavy physical exercise and the onset of paralysis from polio.[7] In 1939 he joined the Children's Hospital Research Foundation of the University of Cincinnati—where he would be his own boss, as he liked to say. Research grants came from the NFIP, later the Public Health Service, and the Surgeon General's Office that would put him in awkward positions (as we will see), but his boast that he always did what he damn well pleased is largely true. In Cincinnati he was free to pursue any research or public health problem.[8]

During World War II Major Sabin worked simultaneously in Cincinnati; at the Rockefeller Research Institute in Princeton, New Jersey; was part of the Commission on Viral and Ricksettsial Diseases (CVR); and went on special missions to vaccinate soldiers or oversee epidemic outbreaks that might affect the military. He performed field tests in several places, including Okinawa, Egypt, the New Jersey State Prison at Trenton, and Longview Hospital in Cincinnati. He seemed to be everywhere.

In 1945 Sabin was promoted to lieutenant colonel and awarded the Legion of Merit for his wartime service. The award recognized his tireless work in combating Japanese B encephalitis, for which he developed a vaccine that was tested and administered to tens of thousands of soldiers during the capture of Okinawa at the end of World War II; for dengue fever research, and sandfly fever prevention.[9] Amazingly Sabin's work also continued on polio, West Nile fever, Rift Valley fever, and St. Louis encephalitis, for which he developed a vaccine. All were serious health threats to the American military.

In addition to the accomplishments cited above, Sabin isolated dengue viruses from Hawaii, New Guinea, and India, and with Walter Schlesinger developed a protective vaccine after the war. And he was the first to isolate several sandfly fever viruses in the Middle East, Sicily, and Naples. Again his special missions were accomplished as he was working in different laboratory situations and advising the AFEB on all manner of epidemic disease. The special missions took place in Egypt (for sandfly fever) in 1943; Okinawa in 1945; Japan (for Japanese B encephalitis) in 1946; Berlin (for polio) in 1947; and Korea in 1951 for a supposed Japanese B outbreak. These last missions occurred after Sabin's active service had ended.[10] That Sabin was able to work successfully on so many fronts during this short period made it no surprise that it took only two years to develop his oral polio vaccine from 1952 to 1954.

Throughout the war, all of Sabin's work was done for the army and under the auspices of the CVR. John Paul of Yale headed this commission, with Sabin acting as chair when Paul was unavailable. The CVR was one specialized committee of the AFEB and answered to Dr. Stanhope Bayne-Jones, a major general of the Army Medical Corps. Sabin was expected to evaluate others' research while deeply involved in his own work. The acerbic Sabin was always full of sharp criticism of work done by others. Joseph Smadel of the Walter Reed Army Hospital urged him to temper his criticisms, because the recipients always took them personally. Even Sabin himself would preface his remarks by saying they were not personal. But he spared no one, friend or foe.[11]

The Sabin style was one of relentless criticism because he felt that was how scientific truth was ultimately attained.[12] Thus was born the universal consensus that Sabin was a hypercritical individual. After the war, while he was working on the relationship of mother's milk to infant polio protection, a frustrating dead end, Sabin was asked to join a semisecret governmental research effort to investigate problems that might result from nuclear and biological warfare and to use his laboratory in Cincinnati for research in that area, things he did not want to do.[13]

To make matters worse, his distractions came in the midst of Jonas Salk's development of an inactivated polio vaccine at the University of Pittsburgh and Hilary Koprowski's early use of a live poliovirus vaccine. He was trumped in 1952 by William Hammon's special gamma-globulin studies that demonstrated protection from polio and Herald Cox of Lederle Labs predicting a huge breakthrough with his oral vaccine.[14] His commitment to finishing his tropical medicine studies, begun during his active service in World War II, had probably cost him his position as the preeminent polio researcher in the world at that time.

With his investigations of dengue and sandfly fever, during and after the war, Sabin needed to do experimental studies, and was reluctant to do so on soldiers because there could be no control group. Instead, prisoner volunteers from Trenton State Prison were chosen for both dengue and sandfly fever experiments. In 1975 he vigorously defended the use of prisoners as necessary and of benefit to the prisoners who regarded it as making a real contribution to humanity.[15] He pointed out that he also used American soldiers and one British colonel for sandfly fever mosquito control tests in Egypt in 1943 because there was no other option. A solution to sandfly fever was vital, he argued, because 70 percent of British forces had been rendered ineffective by the fever in Rommel's first successful attack on Tobruk. Worse, an Italian variant of sandfly fever was prevalent in Sicily as American troops prepared to invade. To quibble years later about using prisoners for tests in light of the existing situations irritated him. Everyone, he said, should have the right to volunteer.[16]

In 1954, as Sabin continued his dengue test on the prisoners, Dr. Colin MacLeod of New York University, now head of the AFEB, raised the issue of Sabin's use of money compensation and certain unspecified privileges as a violation of the American Medical Association's resolution of December 1952 on the matter of using volunteers for experimental vaccinations.[17] Sabin was unmoved by this sort of worry, as he proved when he tested his experimental polio vaccines at the Federal Reformatory in Chillicothe, Ohio, in late 1954 and 1955. The prisoners received twenty-five dollars per oral dose and three days of good time; Sabin himself received only fifty dollars per diem when he was in Washington advising the AFEB or the National Institutes of Health (NIH) in the 1950s.

The preferred test group for polio, it should be pointed out, was institutionalized physically and emotionally handicapped children. Both Salk and Koprowski used such children, and Sabin tried, but could find no favorable institution. That is why he turned to the younger Ohio prisoners. The use of handicapped children was preferred because polio principally attacked children, and these children could be tested thoroughly, and necessary follow-up studies could be done because they were institutionalized. The situation was likewise for prisoners, and it was common practice. Dr. John Fox of Tulane University used 125 prisoners in Mississippi to test epidemic typhus vaccine.[18] Dr. Paul himself had used prisoners in Danbury, Connecticut, to test infectious hepatitis.[19] He came down with it himself and later wondered about the use of human guinea pigs. Dr. Robert Ward of Willowbrook Hospital in New York (and a member of the Medical Corps) used Mexican agricultural workers in Kern County, California, to test western equine encephalitis.[20] And Walter Schlesinger, who developed a dengue fever vaccine with Sabin, used seventeen Connecticut prisoners to test the vaccine.[21]

Sabin did not mention an even thornier problem. His desire to isolate sandfly fever led him to move quickly to undertake tests in humans. His superior, Colonel Bayne-Jones, wrote in 1943 that "it was exciting to hear that Major Sabin had success in transmitting Sicilian fever. Major Sabin is to be congratulated on the speed and efficiency with which he made arrangements for human experiments at Longview State Hospital in Cincinnati."[22] Longview State Hospital was a large facility exclusively for the mentally ill. The use of the mentally ill continued for sandfly fever and dengue fever at least until 1954 and involved about a dozen patients.[23] Transmitting the virus to patients allowed Sabin to confirm the fundamental properties of the virus itself, the first step in developing a viable vaccine.

The conquest of epidemic disease was an important issue of the armed forces after the war and they responded with large research grants. The Virus and Rickettsial Disease Commission alone supplied $161,457.79 in research grants in 1953 and 1954. Sabin received $21,250 to continue his Japanese B encephalitis, sandfly fever, and dengue work. John Enders of Harvard, a Nobel Prize winner for polio research, received $7,810.24 for his work on measles, which yielded a successful vaccine now used universally in young children. John Paul received $58,141.06 for Yale to study viral hepatitis and polio as a military disease. Hammon received $35,269.49 for his work on all the arthropod-borne viruses, including Japanese B, and St. Louis eastern equine encephalitis. German measles and typhus were also diseases that merited research money. Much of the research was a continuation of work begun for the armed forces during World War II.[24]

Sabin's distinguished army service and continuing role as an adviser to the AFEB placed him in a difficult political situation in the early 1950s.

Sabin was a Cold War liberal, but was an opponent of nuclear and biological weaponry: He apparently gave donations to Albert Einstein's Emergency Committee of Atomic Scientists in 1946, as Einstein's eloquent letter suggested.[25] And he deplored the House Un-American Activities Committee's accusations against scientists.[26] In spite of his political views, he did continue to undertake his special missions. His clearly liberal political stances were not an issue in the immediate postwar years when he was sent to Japan, Berlin, and South Korea.

Sabin received a most interesting letter from Dr. Dorland J. Davis, of the National Institutes of Health, on 10 July 1951, informing him that the Public Health Service was organizing a Section Research Program for Biological War Defense with the NIH. "The objectives of this program," Davis said, "are to encourage original research by independent investigators on diseases that may be important in biological warfare defense and to expand and strengthen existing civilian laboratories for specific diagnosis of outbreaks of disease not ordinarily included in routine laboratory techniques." The country would be divided into twelve regions selected with the approval of the National Advisory Health Council and the Influenza Study Program to coordinate the research. Jonas Salk was to be the coordinator of the Pennsylvania, western Ohio, Kentucky, and West Virginia section. Each section was more or less on its own.[27]

Salk assured Sabin that the program was to foster research and only secondarily might research laboratories be asked to do anything else. Salk later learned, in October 1951, that the program would urge the study of many forms of encephalitis, epidemic typhus, Q fever, plague, Rocky Mountain spotted fever, anthrax, botulism, brucellosis, tularemia, and psittacosis. Sabin was still not clear about the exact purpose of the program, but he guessed it was to perform diagnostic tests for those diseases. He evasively told Salk he could not do these sorts of tests in his laboratory at this time. When pressed in 1952 he said he was heavily involved in his polio work. Polio was not one of the suggested diseases to investigate.[28]

In January 1952 the Section Research Program in Microbiology announced that ninety-eight laboratories across the country were cooperating. Sabin acquiesced and listed neurotropic viruses and dengue fever as his field of individual interest; Salk listed poliomyelitis and influenza; Enders, Paul, and Cox at Lederle Labs, Richard Shope at Merck and the Rockefeller Institute, and William Hammon agreed to make their labs available if needed. At least on paper, Sabin would cooperate.[29]

The research that was produced was printed in two thick volumes directed and printed by the Surgeon General's Office, one in September 1953 and the other in December 1954. Copies were given only to the scientists involved. Sabin's contributions were a basic summary of his Army related work: Japanese B encephalitis, dengue, and sandfly fever. Salk did a report on influenza immunization. Hammon reported on encephalitis. It seemed, at least with Sabin's contributions, that the reports were simply of ongoing research and of little value or relevance to biological warfare.

These volumes contained numerous scholarly papers of research relating to nuclear warfare as well as biological warfare. The whole project involved hundreds of scholars and covered such topics as biological and medical aspects of ionizing radiation, radiation and thermal burns, and environmental hygiene. Besides the luminaries in virology, Dr. Michael DeBakey, the heart surgeon, submitted "Factors Influencing Survival in Experimental Penetrating Wounds of the Heart." (The subjects were dogs.) "Neuropsychiatry and Stress" was contributed by Hans Selye, and Fred Charles Iklé, later to explore the same topic for the Reagan administration, wrote on the evacuation of populations in the event of nuclear war.[30]

All of this must have struck a discordant note with the liberal Sabin. He was invited to every Pugwash gathering for years starting in the late 1950s (he did attend in 1963),[31] and he wrote a short article in the *Bulletin of Atomic Scientists* in 1962.[32] Whatever misgivings Sabin may have felt about the Section Research Project, the project itself did not interfere with Sabin's most productive period of polio research, and certainly guaranteed that other research money would come to him. By early 1955 Sabin had selected his three avirulent polio vaccine strains and tested them on his own children and the prisoners at Chillicothe. He needed large field tests, but was stymied. The licensure of the Salk vaccine in April 1955 and Sabin's perceived anti-Salk public stance made a field test in the United States unlikely.

In February 1956 a small group of Russian virologists wanted to visit the United States and talk with prominent American researches at their laboratories. Sabin was asked to write a report on the conversations with the visitors, which he did for the Public Health Service, the Surgeon General's Office, and the Central Intelligence Agency.[33] The report was about Dr. Mikhail Chumakov; his wife, the medical researcher Marina Vorshilova; and Dr. A. A. Smorodintsev. Smorodintsev had worked with Sabin in World War II, and Sabin had advised Smorodintsev on Japanese B encephalitis vaccination in May 1944 in a memorandum labeled SECRET. Because Sabin knew him as an able colleague, it must have amused him when he read a character analysis of the Russians (presumably from the CIA) that described Smorodintsev as a "fanatic communist whose face lights up only when he is discussing viruses."[34] The Russians, especially Chumakov, were greatly impressed by Sabin's work, and invited him to Leningrad and Moscow. An old World War II connection had paid clear dividends.[35]

Sabin's World War II boss, Gen. Stanhope Bayne-Jones, now at the Surgeon General's Office, told Sabin "we are all in favor of your accepting such an invitation."[36] Sabin would oversee the production of his vaccine for trials. He would provide candidate vaccine viruses for study and write his observations as he saw necessary. No military, State Department, or CIA official seems to have thought Sabin was less than reliable because of his liberal politics or antinuclear beliefs. Even if his personal views were known, his wartime service and continued service to the AFEB and the NIH made him above reproach. He traveled several times to the USSR to oversee the production of his oral polio vaccine and swap viruses with the Russians, as the NIH and CIA thought necessary.

By 1960 the Sabin vaccines had been administered to eighty million Russians and twenty-three million Eastern Europeans, virtually wiping out polio. Sabin used the Russian and Eastern European results to push through the field trials and licensing of his vaccines in the United States. Sabin's praise for Chumakov, who directed the Soviet vaccinations, was to give him the title General Chumakov, his highest compliment.[37] The skepticism of American scientists, who distrusted Russians as having no respect for human life, was overwhelmed by the credible results.[38] It is impossible to think of another instance of such close American-Russian scientific cooperation in the 1950s. And it is difficult to think Sabin could have been as free to travel and work with Soviet scientists without his army experiences, and his advisory role to the Surgeon General, NIH, and the AFEB. Moreover, the Russians took a chance with Sabin. If something had gone wrong in the mass trials (they were really mass vaccinations, not trials), Sabin's reputation would have been ruined. Neither the Public Health Service nor the National Science Foundation sponsored him openly for the Russian trials. And by a turn of fate, the great success of the Sabin vaccines led some in the world to think of it as a Russian vaccine.[39] Sabin, an immigrant born in Russian-occupied Poland, was even depicted in *Time* magazine as being Russian or a Russian speaker (he was not). The peripatetic Sabin quickly corrected those mistaken notions: This was a triumph of American science, with Russian cooperation. It was to be Sabin's proud assertion, however, that the superpowers should act by the example he had set with Chumakov.[40]

Sabin did not need his army or government connections to get himself embroiled in the Cold War. In October 1962 he was invited to attend the December meeting of the National Medical Congress in Havana. Sabin always consulted with local medical authorities on their polio eradication campaigns, and to see that the production of the vaccine met his standards. The Russians had given Sabin's vaccine to Cuba. (Incidentally, Sabin always gave the vaccine rights away for free, but he demanded control over the manufacturing process.) He wrote to Walter Rudolph, who handled International Scientific Affairs for the State Department, saying that if the State Department thought it contrary to the best interests of the United States, he would decline. He acknowledged there might be propaganda aspects to his visit. He said he would go "if it would perhaps even be of value to the present situation if I accept the invitation." He declined the invitation the next day.[41]

In 1967, when Sabin did meet Cuban president Fidel Castro, he was pleased that Castro praised his antinuclear war sentiments. Sabin also learned that Revolutionary Defense Committees were established in 1962 to protect Cuban cities, block by block, in the event of an American invasion. After the missile crisis ended, these committees were used to keep track of unvaccinated children and make certain they received their Sabin oral polio vaccine. Thus, by 1967 these committees helped to nearly eradicate polio in Cuba, much to Sabin's pleasure.[42] This was the kind of aggressive public health approach Sabin admired for polio, and it was turning swords into plowshares.

Sabin's service in World War II and his links to the army were not only the source of many significant scientific achievements, but also contributing factors to the Sabin style, a style that demanded leadership on scientific questions, that left no public health issue unanswered, and that required hard work and dedication. It was a no-nonsense military style that, in a celebrity-saturated world, was usually portrayed negatively, giving Sabin a sinister edge compared to the charming Salk. *Time* characterized Sabin as acidulous (that is, slightly sour and aggressive).[43] Jane Smith, a Salk biographer, describes Sabin as full of jealousy, a term also used by Basil O'Connor, head of the NFIP, against Sabin over his criticisms of the Salk vaccine in 1955.[44] Tony Gould, in *A Summer Plague*, refers to him as wily and accuses him of stage managing his attack against Salk, even strangely calling him "Montgomery rather than Rommel or Patton."[45] Some of his assistants in Cincinnati joked that you should watch out for the cane the elderly Sabin brandished, as they pointed to dents in the wall from Sabin banging their heads for mistakes they made at times.[46] It seemed the public perception of his character and personality seemed more important than the scientific issues at stake.

To Sabin it was about leadership on the public health issues. Just as he took leadership in the whole polio drama, he expected others to do their duty. He blamed Surgeon General Leroy E. Burney for needless deaths resulting from delays in licensing his vaccine.[47] He praised the leadership of the local health societies that organized, without the help of the Public Health Service, the great mass immunizations of one-hundred million Americans from 1962 to 1965. He demanded the vaccine be given free and that everyone in the world, not just the United States, should have it.[48] He criticized the World Health Organization for misuse of his oral vaccine.

Leadership got important things done, or it was fair game for criticism. Indeed, Sabin convinced Rotary International to donate millions to help educate polio immunization. In 1976 Sabin broke ranks with the Public Health Service and strongly urged that national mass vaccinations not take place. He was relentless in his criticism of official ineptitude (especially of David

Senser, director of the Centers for Disease Control) and said, "A vaccine is a weapon and it should aimed at the proper target at the proper time; we shouldn't shoot until we see the whites of their eyes, so to speak."[49] This was the leadership he had shown in the large vaccine campaigns against Japanese B encephalitis in World War II and against polio seemingly everyplace on earth. It was a leadership that warned in 1992 that an AIDS vaccine would be impossible to develop in any reasonable period of time.

The relationship of one of the world's great scientists with the armed forces was fruitful and long lasting, and certainly benefited both parties. In 1976 retired general Mark Clark thanked Sabin for sending him the regards of the journalist Lowell Thomas, and added, "I have known of your great service to country, and in particular, your splendid assistance to the military while we were carrying out our various missions all over the world in World War II. I certainly hope our paths will cross soon, for it would be an honor to meet you. I'll work it out and look forward to shaking your hand."[50] The gratitude of the army to Sabin was obvious. He showed his to the army by being buried in Arlington National Cemetery in 1993.

NOTES

1. "Death Throes of a Crippler," *New York Times*, 27 February 2003, A26. The focus is now on raising enough money to eradicate poliomyelitis. See also Guatam Naix, "WHO Shifts Its Goal on Polio to Eradication," *Wall Street Journal*, 15 February 2003, 27. The rise in new cases from six hundred to eighteen hundred in 2002 was a step backward in the eradication campaign.

2. John Paul, *A History of Poliomyelitis* (New Haven, CT: Yale University Press, 1971), 423. Paul was a distinguished polio researcher at the Yale University School of Public Health, and a friend of Sabin.

3. Sabin produced more than 400 scientific papers in his medical research career. Twenty percent were on tropical fevers.

4. Sabin to Mikhail Chumakov, 22 January 1960 (Albert B. Sabin Papers, File Individual Correspondence M. P. Chumakov, 1956–1960, University of Cincinnati Medical Heritage Center, Cincinnati, OH). Sabin praised Chumakov repeatedly as General Chumakov. See also Saul Benison, "International Medical Cooperation: Dr. Albert B. Sabin, Live Poliovirus Vaccine and the Soviets," *Bulletin of the History of Medicine* 56 (1982): 460–85.

5. Sabin to Paul, 31 December 1942 (Sabin Papers Military Service, AEB/NVC Correspondence, 1941–1947). See also Maggie Yax, "Beyond Polio: Albert B. Sabin, World War II and Tropical Medicine," *Ohio Archivist* (Spring 2000): 3.

6. Dr. Saul Krugman to the Felugi Prize Committee (Sabin Papers, File Miscellaneous, Fiuggi Prize Nominations, 1987). Krugman had worked with Sabin on many projects in New York in the 1940s and 1950s, and was, incidentally, his cousin.

7. Milton Silverman, "Exercise and Polio," *San Francisco Chronicle,* 22 April 1948, 1. There was possibly an armed truce between the virus and the patient when paralysis did not occur. This is a typical Sabin metaphor.

8. Sabin to Paul, 29 September 1945 (Sabin Papers, File Military Service, AEB/NVC Correspondence, 1941–1947). Still, Sabin had doubts about his future research projects or if he wanted to stay in Cincinnati.

9. Wolfgang Saxon, "Dr. R. Walter Schlesinger, 89; Developed Dengue Fever Vaccine," Obituary, *New York Times,* 20 January 2003, A21. Sabin's Legion of Merit Award lists his medical accomplishments for the army.

10. Sabin listed his special missions on his vita as Panama, Korea, Japan, China, Germany; he omitted or forgot the Philippines and Egypt.

11. Sabin to Hammon, 18 October 1951 (Sabin Papers, File Military Service, AEB/NVC Correspondence, 1941–1947).

12. Sabin's criticisms were sometimes seen by others as motivated by professional jealousy or simply excessive. See Joseph Smadel to Sabin, 20 October 1951 (Sabin Papers, File, Individual Correspondence). Basil O'Connor of the March of Dimes leveled the jealousy charge and Sabin defended his critical attitudes as did Dr. Hart van Riper of the same organization: Hart E. van Riper to Sabin, 11 May 1954 (Sabin Papers, File NFIP 1940–1961).

13. Sabin to Jonas E. Salk, 22 October 1951 (Sabin Papers, File Section Research Program in Microbiology, Correspondence). He thought one of his projects might be in line with the program, but he was unwilling to turn his laboratory over to the program.

14. See Howard Howe to Sabin, 29 December 1952 (Sabin Papers, File NFIP 1940–1961). Howe told Sabin the National Science Foundation was fully committed to Salk's inactive vaccine, and to go along with the organization. Sabin dismissed Hammon's work as impractical; see Sabin to AFEB, 27 June 1951 (Sabin Papers, File Military Service, Correspondence 1947–1954). Cox's was called misleading: see "Doctor Sabin of Cincinnati Takes a Dim View," *Cincinnati Post and Times Star,* 11 October 1952, 1. Koprowski, Sabin's main rival in developing an oral polio vaccine, recounted his work and Sabin's reaction in "Poliomyelitis: Visit to Ancient History," in *Microbe Hunters: Then and Now,* ed. Hilary Koprowski and Michael A. Oldstone (Bloomington, IN: Medi-Ed Press, 1966), 149 ff.

15. Sabin, "Military Prisoners and Other Adult Volunteers in Medical Research: Benefits to Society and Volunteers," address to Academy Forum (N.A.S, USA), 19 February 1975.

16. Sabin, "Military Prisoners."

17. Colin MacLeod to Paul, 17 November 1954 (copy in Sabin Papers, File Military Service, AEB/NVC Correspondence, 1941–1947).

18. John Fox to Colin MacLeod, 27 June 1952 (copy in Sabin Papers, File Military Service, AEB/NVC Correspondence, 1948–1954).

19. Paul to Sabin, 15 December 1943 (Sabin Papers, File Military Service, AEB/NVC Correspondence, 1941–1947).

20. Robert Ward to Sabin, 4 August 1943 (Sabin Papers, File Military Service, AEB/NVC Correspondence, 1941–1947)

21. Saul to Col. Adam Rapalski, 5 November 1954 (copy in Sabin Papers, File Military Service, AEB/NVC Correspondence, 1947–1954).

22. Stanhope Bayne-Jones to Paul, 21 October 1943 (Sabin Papers, File Military Service, AEB/NVC Correspondence, 1941–1947).

23. Sabin to Paul, 22 November 1952, 27 April 1954, and 26 July 1954 (Sabin Papers, File Military Service, AEB/NVC Correspondence, 1948–1954).

24. Commission on Virus and Rickettsial Diseases, AFEB: Proposed Studies and Budgets for 1953–1954 (Sabin Papers, File Military Service, AEB/NVC Correspondence, 1948–1954).

25. Albert Einstein to Sabin, 6 August 1947 (Sabin Papers, File Miscellaneous Unsorted).

26. Sabin to Rep. Charles H. Elston, 15 May 1948 (Sabin Papers, File Unsorted Correspondence).

27. Dr. Dorland Davis to Sabin, 10 July 1951 (Sabin Papers, File United States Public Health Service, Section Research Program in Microbiology, 1951–1952).

28. Jonas Salk to Sabin, 24 August 1951 (Sabin Papers, File United States Public Health Service, Section Research Program in Microbiology, 1951–1952). See also Salk's letter of 13 October 1951, and Sabin's irritated response of 22 October 1951.

29. Research Medical Report, Army Medical Service, 1st Quarterly Report, 1 July–30 September 1953, Research and Development Division Office of the Surgeon General, U.S. Army.

30. Research Medical Report, Army Medical Service, 1st Quarterly Report, 1 July–30 September 1953, Research and Development Division Office of the Surgeon General, U.S. Army. A second volume, now an annual report, followed for 1954. The program ended after 1954.

31. "Pugwash Conferences, 1958–1970," Sabin papers, File: "Miscellaneous."

32. Sabin, "Breaking the Impasse," *Bulletin of Atomic Scientists* (October 1963): 35.

33. Sabin, "Recent Work on Certain Virus Diseases in the USSR. Highlights of discussions with professors M. P. Chumakov and A. A. Smorodintsev during their visit to Cincinnati, February 3–5, 1956." 27 February 1956 (Sabin Papers, File USPHS, USSR Visitors, 1955–1956).

34. Memorandum on Discussions with Dr. A. A. Smorodintsev, May 1944 (Sabin Papers. Unsorted).

35. H. Van Zile Hyde to Sabin, 13 Jan. 1956 (Sabin Papers, File USPHS, USSR Visitors, 1955–1956).

36. Bayne-Jones to Sabin, 20 February 1956 (Sabin Papers, File USPHS, USSR Visitors, 1955–1956).

37. See n. 4, above.

38. Benison, "International Medical Cooperation," 480. Stanley Plotkin, the developer of the rubella vaccine, still finds the Russian results literally unbelievable. See Edward Hooper, *The River: A Journey to the Source of HIV and AIDS* (New York: Little, Brown, 1999), 739.

39. Sharon Moloney, "Russians Getting Credited for Polio Vaccine—Sabin," *Cincinnati Post and Times Star*, 20 October 1961, A20.

40. Sabin, "The Role of My Cooperation with Soviet Scientists," *Perspectives in Biology and Medicine* 31 (1987): 57–64. According to Dr. Chanock, Sabin did not trust Smorodinstsev. In 1960 the Soviet Academy of Sciences reprimanded Smorodinstsev for trying to pass off some of Sabin's work as his own (Robert Chanock, personal letter, March 2004).

41. Sabin to Walter M. Rudolph, 4 October 1962 (Sabin Papers, File Individual Correspondence: Walter M. Rudolph).

42. Sabin to Chumakov, 16 September 1978 (Sabin Papers, File Individual Correspondence, 1974–1992).

43. "Polio Imbroglio," *Time*, 3 February 1961, 50.

44. Jane S. Smith, *Patenting the Sun: Polio and the Salk Vaccine* (New York: William Morrow, 1990), 300–305.

45. Tony Gould, *A Summer Plague: Polio and Its Survivors* (New Haven, CT: Yale University Press, 1995), 153.

46. Lew Moores, "Dr. Albert Sabin, the Man Who Conquered Polio," *Cincinnati Post*, 23 April 1990, 1, p.4A.

47. Sabin, Letter, *Barrons*, 16 January 1961, 7.

48. Sabin to Alexander Langmuir, 11 January 1960 (Sabin Papers, File Oral Polio Miscellaneous Correspondence, 1954–1961).

49. "Drs. Salk, Sabin Disagree on When to Use Swine Influenza Vaccine [1976]," See also "Swine Flu: Specter of 1918 and Dilemma of 1976," *Cincinnati Horizons* 6 (October 1976): 12–15; "Dr. Sabin Blasts Swine Flu Policy," *Spartanburg Herald*, 14 September 1976, 2B.

50. Mark Clark to Sabin, 4 March 1976 (Sabin Papers, File Miscellaneous, Individual Correspondence: Mark Clark).

8

A New R&D Order: The Rise of Big Engineering in the Second World War

Robert G. Ferguson

World War II was a pivotal juncture in the evolution of military-sponsored research and development in the United States. The Manhattan Engineering District, the development of radar and other advanced electronics, and the creation and production of antibiotics stand as precursors to the country's Cold War R&D efforts. Scholars have analyzed postwar R&D patterns, especially the spectacular rise of particular scientific disciplines, the development of revolutionary theories and technologies, and the political and economic foundations that nurtured these efforts. Big Science, the Endless Frontier, and the military-industrial complex are part of a rich discourse characterizing these fundamental shifts and emphases.[1]

Scholars have paid far less attention to the impact of the war on industrial R&D. This chapter[2] focuses on the aircraft industry, later to become the archetypal corporate actor in the Cold War drama. The war profoundly altered engineering capabilities, changes that survived the abrupt halt to production in August 1945.[3] Continuous aircraft development had become irrevocably dispersed among many firms and across new disciplines, with consequent changes to firm organization and technology. Nor did engineers and managers forget what they learned during the war. Aircraft production, in short, no less than the A-bomb project and other better-studied wartime programs, helped lay the foundations for a research and development order in the later twentieth century.

One of the keys to understanding R&D in the aircraft industry is that development encompassed both design and production phases. The basic reason is simple. How a new aircraft performed remained a question mark until it underwent testing. More complex reasons included the contribution of production to aircraft design, as well as the continuous stream of change requests from the customer. These factors resulted in a very tight coupling between design and production with high levels of social control over engineering knowledge (the what, the how, and when of manufacture). It is this control over engineering knowledge that unites the development process and serves as the focus for this investigation. Knowledge in an aircraft company was expressed in a variety of ways, including, for example, contract specifications, wind tunnel data, blueprints, parts cards, shop orders, templates, control gauges, worker techniques, flight test results, and change orders. World War II was not merely a period of vast production and advanced aircraft designs; it presented new challenges to the creation and transmission of knowledge. This chapter outlines four categories of analysis for understanding these problems: subcontracting and dispersed production, collaboration, concurrent design and manufacture, and complexity.

SUBCONTRACTING/DISPERSED PRODUCTION

The tendency of the aircraft industry in the interwar period, and indeed in the immediate postwar era, was to eschew outside production. There were, of course, various items that the manufacturers regularly purchased, including off-the-shelf parts such as screws and cables and highly specialized items such as engines and instruments, but between these extremes manufacturers coveted whatever work they could garner for their modest factories. This predilection changed only when firms found themselves with more aircraft orders than they could absorb.[4] They had good reasons for avoiding outside manufacture. Outside manufacture obviously incurred new charges, such as transportation and packing fees. And in lean times, which the 1930s and postwar years most certainly were, keeping production in-house was sometimes the difference between survival and bankruptcy. However—and crucially—outside production took processes outside of the knowledge ecology of the main plant. Where aircraft manufacturers relied on efficient and informal communication between engineers and shop workers, and most of these firms did in the 1930s, dispersed production required formal substitutes for the communication of engineering knowledge and the recreation of such shop-floor knowledge as practices, special tooling, and embedded norms. Widespread aversion to subcontracting only exacerbated these knowledge asymmetries.[5] Except for a handful of companies that had deep knowledge in a particular component, like propellers, or in areas of high geographic clustering such as Southern California, qualified aviation subcontractors were few and far between, especially when compared to their counterparts in the automotive industry.

The movement to subcontracting and its in-house equivalent, feeder/branch plants, grew dramatically with mobilization. Boeing provides a good example. Before the war, the company did not subcontract any portion of its early B-17 bomber designs (Model 299 to B-17B). By 1944 the company was subcontracting more than 32 percent of the aircraft (by man-hours). In addition, where production was formerly self-contained at the company's two Seattle plants (located very near each other), during the war they chose to spread fabrication across a network of seven Boeing-owned and -managed feeder plants in the state of Washington.[6] Other companies experienced similar shifts, all compounded by the dispersal of assembly operations to duplicate factories in the country's interior. In Boeing's case, final assembly for its large aircraft took place in Seattle; Renton, Washington; and Wichita, Kansas. While Boeing had operated the Wichita facility for a number of years, this particular factory had manufactured a distinct product line, the Stearman biplane. With the war, assembly of the same types took place in multiple locations.

Mobilization highlighted the aircraft manufacturer's lack of experience in arranging and managing outside production. Among the important lessons of the war was the need to exert greater control over the subcontracting process. But even in cases where subcontracts went to well-regarded engineering organizations (including the prime contractor's own remote facilities), manufacturers learned the hard way just how difficult it was to transfer engineering knowledge, such as blueprints or templates, between locations. Specifications would go out, but the subcontractor or remote plant would return a product that did not fit or pass inspection.[7] Part of the problem lay in the fact that aircraft designs were socially and physically embedded within a particular factory. Drawings and work orders were tied to a firm's unique system for creating and transmitting engineering knowledge. While other manufacturers had somewhat similar methods for design and production—most, for instance, used parts cards and work orders—these were not standardized. And while a drawing of a wing would be recognizable as such among different companies, exact interpretations would differ, an unacceptable outcome for a technology that demanded close tolerances in order to operate. Such differences extended well beyond paper representations of engineering knowledge. Various companies employed the same types of machine tools and forming equipment but evolved different procedures and methodologies for using them. Some of these differences fall into the category of tacit knowledge, a well-known pitfall of technology transfer, but many differences were explicitly codified in company and shop manuals.[8] And all of this assumes that the aircraft drawings were complete in the first place. Design engineers routinely waited until after workers completed the first prototypes in order to make complete drawings. The determination of wiring pathways, for example, was a job best left to the skilled workman. What were prime contractors and subcontractors to do?

Manufacturers learned that sending blueprints was rarely enough. Firms did not transfer only designs, but whole administrative and factory processes. The elements that made the home factory work in the first place had to be re-created at the remote site. In addition to the paperwork, which included blueprints, parts cards, process charts, and production illustrations, went jigs, templates, unique tooling, gauges, and master gauges. Firms sent sample parts, sample assemblies, and sample aircraft. And they exchanged workers, letting outsiders observe processes at the prime contractor's plant, and placing experienced workers inside the remote plant.

These were not simply turnkey operations; the dynamic nature of the product, the difficulty of acquiring strategic materials and machine tools, and the coordination of accelerating and slipping master schedules meant that subcontractors had to be integrated under the prime contractor's managerial umbrella. This placed great strain on management because—and this is very telling—industry practice was to promote from within.[9] Firms did not seek to grow their managerial ranks through individuals unfamiliar with aircraft industry routines. In most cases, middle management comprised engineers who had already put in time at the drafting tables, drawing engineering details and calculating aircraft performance. Faced with severe managerial shortages in mobilization, the industry did not change this practice. Aircraft companies functioned because management understood the underlying design and manufacturing processes that were unique to the industry and unique to each firm. With the entrance of subcontractors from outside the aircraft industry, notably the automotive industry, knowledge asymmetries between the two industries would prove most conspicuous.

COLLABORATION

Collaboration invited all the challenges of subcontracting, as well as fundamental political conflicts. Who had authority over design and manufacturing knowledge? World War II saw the production of the same designs by many different companies, some of them outside of the aircraft industry. How were the manufacturers to resolve the inevitable differences among factories, especially when the customer—the military—expected standardized products with interchangeable parts? These hurdles prompted organizational means for identifying and resolving differences, and technological approaches that isolated divergent factory practices. The war also created the opportunity for collaboration outside of production relationships. The manufacturers established cooperative associations that sought to leverage their collective engineering capabilities, partially by sharing proprietary knowledge, but also through new, coordinated research programs that addressed short- and medium-term design and manufacturing problems.

Among all the collaborative programs, the Boeing-Douglas-Vega (BDV) Committee stands out because it established an early, successful precedent for subsequent efforts. In April 1941 the Army Air Corps (later renamed the Army Air Forces) invited Douglas and Vega to join Boeing in the production of B-17 bombers. Boeing, which had begun the B-17 project in 1934, anticipated volume production with the E model, due out in the summer of 1941. The Douglas Company agreed to produce 600 B-17s at a maximum production rate of 60 per month. They would ultimately deliver 3,000 aircraft before ending production in June 1945.[10] Vega initially agreed to build 500 bombers, but produced 2,750 before the war's end.[11] The three companies and the army set about tackling what everyone realized would be a difficult endeavor. Boeing president Phil Johnson suggested to Gen. "Hap" Arnold that they establish a means for coordinating production as well as technology transfer. In May they created the BDV Committee, an official arm of the Army Air Corps Material Division, chaired by Brig. Gen. K. B. Wolfe. In addition to the executive committee were subcommittees on tooling, engineering, and procurement. Overall, the BDV coordinated the subcontracting program, controlled and released engineering drawings (including change orders), oversaw production schedules, and helped arrange material and machine tool procurement. Like other wartime collaborative programs, the BDV did not concern itself with fiscal matters, leaving these for the military and its respective contractors to resolve.[12]

Anticipating differences in engineering and production cultures, the BDV Committee established an office in Seattle where employees from the three companies and the military could oversee joint operations. At the committee's height, more than two hundred people worked in the BDV office. One of the first obstacles they confronted was the release of some six thousand drawings. Not only were they incomplete, but Boeing's methods of engineering and production communication (drawings, parts cards, bills of material, etc.) differed sufficiently from Douglas and Vega so as to be technically unintelligible. The Southern California companies each provided fifteen engineers to assist with the drawings and upon their return from Seattle, these engineers acted as interpreters at their home plants.[13] Similarly, Douglas and Vega cooperated in the development of tooling drawings, something with which Douglas had significant experience. Indeed, Boeing's Tooling Division dated only to the year before.[14]

While Boeing, the prime contractor, retained ultimate authority over the B-17's design, the BDV subcommittees provided a negotiating space to communicate problems and arbitrate solutions. With three different manufacturing methods and three different traditions of factory practice, the BDV faced significant coordination challenges. One expedient that proved important for the future was establishing levels of interchangeability. Specific parts of the aircraft, such as the wing assemblies, had to be interchangeable in order to meet the army's specifications. Within the bounds of these points of interchangeability, manufacturers were free to modify smaller details of design so long as it did not impact strength or weight.[15] This effectively reduced the challenge of commonality and quarantined the entropic effects of idiosyncratic firm practices. As it did with so many aspects of the BDV program, Boeing used the B-29 Committee to refine these practices.[16]

The war was a watershed for initiating widespread cooperative activities among the manufacturers. The broadest efforts would be found in two groups: the Aircraft War Production Council (AWPC) and the AWPC East Coast. While never completely isolated from each other before the war, the manufacturers generated much of their engineering information, especially production information, in-house. The principal exception to this was wind tunnel data, the province of NACA wind tunnels and university aeronautics departments. The challenge for the councils was to establish means for not only sharing information but also generating it cooperatively. Encouraging disparate groups to exploit outside information led to special efforts to tailor information retrieval to the likes of engineers. Mutual engineering research, like cooperative aircraft production schemes, would eventually take on the methodology of standardization in order to make research results portable.[17]

CONCURRENT DESIGN AND MANUFACTURE

Concurrent design and manufacture became one of the most destabilizing of the war's influences, because it meant that designs and factory operations were never fixed. The R&D portions of the process could not function in isolation, and the production portions could never find any stability that would permit Fordist mass production. These symptoms stemmed from a single problem: The military demanded advanced products very quickly. For aircraft that were already in production, this meant incorporating a nonstop stream of changes learned from flight-testing and battlefield experience. For entirely new aircraft designs, this meant rushing the aircraft into production well before developmental issues were resolved, as exemplified by the case of the B-29. Concurrency, however, was not simply a matter of parallel design and manufacturing activities. It involved a constant reiteration of the development process, including the reengineering of designs that were already tooled on the factory floor. By way of example, production of both the B-25 and P-51 began in 1941, yet engineering activities did not peak until 1943. For the Boeing B-29, the ratio of engineering man-hours in the production phase was nearly ten times the man hours required for the original design.[18]

Concurrency, of course, rested on the ability to incorporate changes into an operating manufacturing line, but it also demanded great agility in reengineering the aircraft on the fly. Changes had to be defined and validated on paper before they could be reflected on the shop floor. This was a rigorous process, since parts and assemblies had to be redrawn; performance recalculated based on changing weight, strength, and other parameters and approved by multiple departments; amendments made to parts cards; and revisions distributed according to a planned and tightly controlled schedule. This is to say nothing of contractual negotiations with the army and navy over costs and deliveries. On the production side, changes demanded new shop orders, new tooling, new materials and parts orders, modified subcontracts, and new work processes. Keeping in mind that aircraft underwent numerous large-scale changes, such as increased armament, each involving hundreds of smaller detail changes, the task was more an organizational morass than it was a tooling challenge. The B-25 and P-51 programs noted above underwent 1,625 and 1,084 Master Change Records (MCRs), respectively. Each MCR was a general grouping of changes that, for North American, required about a thousand hours of engineering work.[19] Even with agile engineering departments and quick tooling changes, manufacturers worked against their factory's "flow time," the period from the fabrication of the first part to the emergence of a finished aircraft, which could be as long as two months. Factories could make immediate changes outside of the normal production process, but this often meant discarding parts and assemblies built weeks earlier. In most cases, companies worked from both ends of the process, pushing engineering change orders through the usual channels while performing quick modifications right on the assembly line.[20]

The continuous development of wartime aircraft placed large burdens on engineering departments. Companies hired what engineers they could find, and some, like Curtiss-Wright, established programs to train female draftspersons and engineers. Organizationally, most companies were at a distinct disadvantage because they were not designed for the speedy flow of engineering knowledge. Before the war, most companies, if not all, employed functional divisions within the engineering department. The weight group would perform weight analysis for all of a company's aircraft, the mechanical group would design all the mechanical features, and so on. Engineers would have found themselves affiliated more with a subdiscipline than with a particular aircraft project. Project engineers oversaw the development of specific aircraft and were assisted by leaders from the various sections, but each project engineer had to work through the bureaucracy of the functional groupings.[21] This structure would have been especially cumbersome for companies working on the development of many aircraft.[22] Engineers within the subgroups would have been less aware of relative priorities among tasks from different projects, and they would have faced greater difficulty resolving problems that stretched across disciplinary lines.

At least one company, Douglas, revolutionized its engineering department in a way that would be termed "matrix management" in the 1950s. For its Long Beach Engineering Department, Douglas used project groupings to create the dominant organizational divisions, while retaining weaker horizontal links to specific engineering groups.[23] The arrangement gave project engineers greater authority, but it also eased the flow of engineering knowledge, bringing together a cross-disciplinary team that could coordinate changes according to the project's needs. At the same time, engineers retained formal access to the specialist groups, thus providing them with the latest developments in their subdisciplines. Douglas's Long Beach experience was not typical. More common was Boeing's effort on the B-29, which shifted resources from the specialist groups toward the project, yet stopped short of establishing formal horizontal linkages between the project engineers and the specialist groups.[24]

One of the key elements tying engineering and production together was the evolution of block changes. From everyone's perspective, it was more efficient to group engineering changes together in packages (blocks) rather than issue changes as they arrived. For the military, blocking eased procurement and logistical workloads. For contractors, blocking simplified the organizational burdens of controlling design changes. The AAF mandated the block system into effect in July of 1942 and established a common nomenclature for all of its contractors.[25] Blocking was not a panacea, however, as adoption was uneven and manufacturers pursued different organizational means for controlling change implementation. In the case of Boeing B-29 production, blocking was not adequately synchronized across the three separate factories assembling the aircraft. When one factory instituted a new block, it fed back into the subcontractors, and then, ready or not, into the other contractors.[26]

North American Aviation's version of blocking was a work in progress throughout the war. The company sought to group major changes at three-month intervals, a period of time that would have allowed the managers to smoothly adjust purchases and subcontracts, as well as move in step with the factory's flow time. In order to oversee the continuous stream of design changes, North American established a change control group that evaluated and scheduled changes. This group was eventually placed within the manufacturing division and integrated with the scheduling and hours control departments underneath a manufacturing coordinator. In effect, the manufacturing department had substantial authority over aircraft development, though the engineering department had the ability to appeal such "downstream" decisions.[27]

One of the difficulties of coordinating design and manufacturing changes was that these companies did not have established methods for overseeing, in real time, the state of a project. Design changes were reflected in thousands of drawings, the release of which was controlled within a draftsman's office. Production was reflected in shop orders, purchase orders, and tooling, all coordinated out of manufacturing departments. As an example of the scale of this planning quagmire, in June 1943 Boeing's B-29 program produced "a weekly average of 1,720 drawings, deviations, [and] production memorand[a]."[28] North American's change control was an organizational step toward resolving the problem. Late in the war, Boeing established graphical methods for portraying manufacturing and tooling development. Boeing managers oversaw purchasing and inventory control by plotting planned versus actual progress. Similarly, the company's tooling production charts showed when and where tools would be made, as well as completed tooling, hours invested, and estimates of future man-hour requirements.[29]

COMPLEXITY

"Complexity" is a relative and imprecise term, and the aircraft designs tend to become more complex in successive generations. What distinguishes the wartime period is the rapid growth in demand for broader and deeper engineering capabilities. The requisite engineering knowledge was no longer the domain of an individual engineer, or even of an aircraft company. In 1943 Claire Egtvedt, a Boeing engineer and executive, stated, "I think as the design [of the B-29] developed[,] it developed more as an organizational airplane than any that had been previously made."[30] Complexity turned design engineering into an organizational task that integrated the work of specialty teams spread across many firms. Complexity elevated subcontractors into design partners, collaborating on designs so tightly linked that engineering changes swirled in a vicious circle, threatening to mire projects in their tracks.

Three technologies contributed significantly to the rise in complexity: pressurization, electronics, and propulsion. Pressurization, the ability to maintain a normal cabin pressure at high altitudes, was a linchpin linking many different engineering disciplines. Boeing had achieved modest results with pressurization before the war, but the B-29 would be the first true proving ground for the technology. Aviation electronics came of age in World War II. The invention of radar is the best known of these developments, but there were others, such as remote fire control, that signaled the industry's movement toward onboard electronic computers. Finally, propulsion followed two different routes, both highly complex. Traditional reciprocating engines grew in size and intricacy as engineers pushed them to greater levels of power at higher altitudes. New jet turbines, though mechanically simpler than reciprocating engines, redefined airframes and their aerodynamics. World War II not only sped the deployment of these three technologies, it provided uncommon levels of funding for the manufacturers to develop a multitude of new engineering research programs. The aircraft industry that emerged from World War II did not simply have faster, bigger, and better aircraft. It had become scientific.

The notion that the process of aircraft design had changed during the war was not necessarily understood or universally accepted, but there was a slow recognition that R&D and a wide network of outside specialties were becoming essential features of aviation. The incorporation of new engineering specialties meant greater in-house research capabilities as well as a stronger reliance on advanced subcontractors. Still, even as complexity pushed the aircraft industry to broaden its activities, it also required an ability to sew such far-flung programs back together into a working aircraft. The integration of advanced subcomponents was not simply a matter of "bolting on" a new piece of equipment. Subsystems were dependent upon each other, and it fell to the prime contractor to negotiate between systems and to test the integrated product. In this, the aircraft industry was prepared to serve at the top of the contracting pyramid; the testing of unpredictable, integrated systems was part of what it had always meant to be an aircraft manufacturer.

Boeing's records vividly illustrate the quantitative and qualitative increase in engineering research. The company's 1940 budget for design and engineering research amounted to $73,145. The bulk of its investment went to developmental aircraft studies, either the Model 341 bomber, predecessor of the B-29, or refinements to the Model 307 Stratoliner, a commercial passenger aircraft. The remaining studies emphasized mechanical and structural testing.[31] For the fiscal year beginning mid-1944, Boeing's research budget was $1,965,800. Not only is this figure twenty-six times larger than three years earlier, but the 1944 budget includes no development engineering. Equally important, where research was divided into two major categories in 1940, engineering and flight test/aerodynamic, by 1944 the company had research programs for most, if not all, of the company's engineering subdisciplines, such as structural, vibration, and power plant.

A rush to build new labs and test rigs accompanied the wartime rise in research funding. Before the war, most aircraft manufacturers had very modest research facilities, relying to a great extent on NACA and university wind tunnels. Bell Aircraft's engineering laboratory was typical of the interwar years;

the company's facilities and activities were strongly reminiscent of the kind of basic materials testing that the railroads and nascent telephone companies had in the latter quarter of the nineteenth century, but with more sophisticated equipment. Much of the laboratory's activities revolved around routine inspection. All castings and critical fittings required X-ray testing, which in 1940 amounted to a few hundred parts per day. The lab was equipped with tensile, compression, impact, and fatigue testing equipment, and could carry out basic chemical analysis.[32] Less typical was Boeing's equipment for high-altitude testing. In 1937, the company began work on its Stratoliner, a design that was to fly at altitudes of twenty thousand feet using a pressurized fuselage. The company built a twelve-foot-long cylindrical "Strato-Chamber" that could simulate air pressures at forty thousand feet and 30 degrees below zero. By March 1942 Boeing's modest collection of test equipment included a wind tunnel, a "Polar-Lab" for cold temperature testing, and a rotating engine test stand to analyze cooling, vibration, fuel flow and airflow.[33] The wind tunnel, which the company built at a cost of over $800,000, was one of the fastest wind tunnels in operation for its size (seven hundred miles per hour through a twelve-foot test section) and it was, by far, the largest in-house wind tunnel.[34] By 1947 Boeing had fifteen separate laboratories, including: propulsion, aerodynamic, mechanical, structural, electrical, acoustical, hydraulic, metallurgical and electronic, as well as six "flying laboratories."[35] Boeing was not alone in the development of new experimental facilities and specialized research teams. North American Aviation built a 327-mile-per-hour tunnel with a 7.75-by-11-foot test section in 1942, and constructed an extremely large and sophisticated high-altitude chamber in 1945 capable of pressure-altitudes of 60,000 feet and a temperature range of –100 to +200 degrees Fahrenheit.[36] Lockheed built its own pressure chamber in 1942.[37] Consolidated-Vultee began building its own aeronautical laboratory in San Diego in January 1945, incorporating a 350-mile-per-hour tunnel with an 8-by-12-foot test section.[38]

The widespread turn toward in-house engineering research was not necessarily a belated discovery of the benefits of science. While these laboratories did conduct some basic research, much work was oriented toward the problems of systems integration. When these companies established electronics laboratories, it was not because they planned to compete with the likes of Honeywell, Bendix, General Electric, or Bell. These laboratories served as knowledge interfaces for more competent subcontractors. Additionally, such labs performed integrated tests that could not be conducted by individual subcontractors. It behooved the prime contractor to be minimally competent in outside specialties in order to troubleshoot new designs and help resolve differences among competing technical priorities.[39] Radar, for example, was a piece of government-furnished equipment that was often installed at postproduction modification centers. Still, in the case of the

B-29, Boeing designed the "first highly-accurate streamlined radar antenna" for the aircraft.[40]

A NEW R&D ORDER

The core argument of this chapter is that scholars need to reevaluate World War II's influence on corporate development capabilities. The war was more than an exercise in volume production, more than the story of a job-shop industry that learned to do it Detroit's way and then gave it all up in August of 1945. World War II was a formative experience, one that gave firms new capacity to collaboratively generate, communicate, and control technological knowledge. This was a very different industry than the one that entered the war, and not solely by virtue of its size and newfound political support. This is not to say that the war established a template for all postwar technological development. The lessons were as often negative as they were positive, but they were influential nonetheless.

The cancellation of military contracts after VJ Day practically shuttered the American aircraft industry. Some of the wartime trends would reverse themselves. Subcontracting, collaboration, and concurrency would take a hiatus as manufacturers consolidated far-flung activities, procurement dollars evaporated, and development cycles lengthened. The Korean War, and the Cold War in general, revived many of the knowledge challenges of World War II. But there appears to be a fundamental difference in how engineers and managers approached development in the 1950s. Where technical and organizational innovations of the 1940s were usually ad hoc responses tied to specific wartime products, innovations of the 1950s were created as general solutions that could be applied to any large-scale development program. Development itself became an object of formal and systematic analysis. This may help explain why some capabilities that surfaced in the 1940s, such as project management and matrix management, were considered novelties in the 1950s.[41] The development process became not only product-neutral in the 1950s but also the province of specialized organizations such as RAND, Ramo Wooldridge, and the Aerospace Corporation. While systems integration is appropriately located in the 1950s, the rise of postwar development capabilities had its roots in World War II.[42]

NOTES

1. Vannevar Bush, *Science, the Endless Frontier* (Washington, DC: Government Printing Office, 1945); Daniel J. Kevles, *The Physicists: The History of a Scientific Community in Modern America* (Cambridge, MA: Harvard University Press, 1987); Peter

Galison and Bruce Hevly, eds., *Big Science: The Growth of Large-Scale Research* (Stanford, CA: Stanford University Press, 1992); Stuart Leslie, *The Cold War and American Science: The Military-Industrial-Academic Complex at MIT and Stanford* (New York: Columbia University Press, 1994); Steven Rosen, comp., *Testing the Theory of the Military-Industrial-Complex* (Lexington, MA: Lexington Books, 1973); Richard Rhodes, *The Making of the Atomic Bomb* (New York: Touchstone Books, 1986); Robert Buderi, *The Invention that Changed the World: How a Small Group of Radar Pioneers Won the Second World War and Launched a Technological Revolution* (New York: Touchstone Books, 1996).

2. Thank you to Stephen Johnson and Erik Conway for their valuable feedback. This chapter was originally presented under the title, "A New R&D Order: World War II and the Creation of Development Networks."

3. Markusen et al. highlight this discontinuity: "[The aircraft] firms remained small and struggling until the growth of commercial aviation in the mid-1930s, some until the defense buildup of the late 1930s. Somehow, almost all of them met the extraordinary managerial challenge of converting to mass production during World War II. Those that flourished under the profits and government financed increases in capacity during World War II were thus much stronger at war's end. But, in retrospect, this growth was small in comparison with the challenge of the postwar period, when businesses had to convert from airplanes to aerospace. That meant an entirely new attitude to R&D, a capacity to think about the future in futuristic terms, and a need to create areas of expertise in the marriage of aviation and electronics." Ann Markusen, Peter Hall, Scott Campbell, and Sabina Deitrick, *The Rise of the Gunbelt: The Military Remapping of Industrial America* (New York: Oxford University Press, 1991), 99. Stephen Johnson is relatively alone in highlighting organizational continuities between the wartime and postwar eras: Stephen B. Johnson, *The Secret of Apollo: Systems Management in American and European Space Programs* (Baltimore: Johns Hopkins University Press, 2002). For characterizations of the wartime experience as one of mass production, see Jonathan Zeitlin, "Flexibility and Mass Production at War," *Technology and Culture* 36 (1995): 46–79; Gerald D. Nash, "The Impact of World War II on the American Economy," and Robert D. Cuff, "Organizing U.S. Aircraft Production for War, 1938–1944: An Experiment in Group Enterprise," both in *World War II and the Transformation of Business Systems*, ed. Jun Sakud and Takao Shiba (Tokyo: University of Tokyo Press, 1993). Within aviation history, the two most influential accounts have been I. B. Holley, *Buying Aircraft: Matèriel Procurement for the Army Air Forces*, vol. 7 of *United States Army in World War II* (Washington, DC: Government Printing Office, 1964); and W. F. Craven and J. L. Cate, eds., *The Army Air Forces in World War II: Men and Planes*, vol. 6 of *The Army Air Forces in World War II* (Chicago: University of Chicago Press, 1955).

4. John S. Day, *Subcontracting Policy in the Airframe Industry* (Norwood, MA: Plimpton Press, 1956), 121; J. H. Kindelberger, "Problems of a Rapid Expansion of Airframe Production in a Future War," in *Elements of American Air Power*, comp. Aircraft Industries Association (Washington, DC: Aircraft Industries Association, 1947), 104.

5. Day, *Subcontracting Policy in the Airframe Industry*, 124.

6. U.S. Army Air Force, Air Materiel Command, "Boeing Aircraft Company, Seattle, Washington, B-17, Production and Construction Analysis," May 1946, 32–34, 58 (Boeing Historical Archives).

7. I have written on some of Boeing's earlier difficulties in this regard in "Airframe Manufacture and Engineering Exchange," in *Atmospheric Flight in the 20th Century*, ed. Peter Galison and Alex Roland (Dordrecht: Kluwer Academic, 2000), 259–85.

8. See especially Walter G. Vincenti, *What Engineers Know and How They Know It: Analytical Studies from Aeronautical History* (Baltimore: Johns Hopkins University Press, 1990).

9. Day, *Subcontracting Policy in the Airframe*, 58; Aircraft Industries Association, *Elements of American Air Power*, 94, 101.

10. U.S. Army Air Forces, Air Material Command, "Production and Construction Analysis: Douglas Aircraft Company, Long Beach, B-17," (hereafter AAF, Douglas Long Beach, B-17), ix, Air Force Historical Research Agency, Maxwell AFB; Boeing Aircraft Company, B-17 Series Historical Study, Chart I (Box 1096, File 7, Boeing Historical Archives).

11. U.S. Army Air Forces, Air Technical Service Command, Logistics Planning Division, "Industrial Planning Project, Construction and Production Analysis, Lockheed, Burbank," 9 (Air University Library, Maxwell AFB).

12. Mandus E. Bridston, "One for All and All for One," Boeing press release, March 1942 (Subject File Manufacturing 1930–1943, Boeing Historical Archives).

13. Mac Short, "Engineering BDV Production of Flying Fortresses," *Automotive and Aviation Industries* 89 (15 October 1943): 17–21, 92–96; AAF, Douglas Long Beach B-17, 3, 6–7, 36, 43.

14. AAF, Douglas, Long Beach B-17, 31.

15. Short, "Engineering BDV Production of Flying Fortresses."

16. See the Committee Meeting Minutes, 2 vols. (Model 345, Boeing Historical Archives). The B-29 Committee included an Executive Committee and, initially, five subcommittees: Engineering, Tooling, Procurement, Spare-parts and Inspection. See U.S. Army Air Forces, "The Superfortress," 20 (Air Force Historical Research Agency, Maxwell AFB).

17. See Robert Ferguson, "Technology and Cooperation in American Aircraft Manufacture During World War II" (PhD diss., University of Minnesota, 1996).

18. U.S. Army Air Forces, "The Superfortress," 14.

19. North American Aviation, Inc., "A Brief History of Operations Immediately Prior to and During World War II," 38 (San Diego Aerospace Museum; also available at the Boeing Historical Archives).

20. U.S. Army Air Forces, Air Technical Service Command, "Boeing Aircraft Company, Seattle, Washington, B-29, Production and Construction Analysis," 21 January 1946 (hereafter AAF Renton), 38–39 (Boeing Historical Archives).

21. Val Hoffman, outline for course in manufacturing procedure based on Boeing practice, October 1939, 40 (box 4704, file 4, Boeing Historical Archives).

22. Guy Vaughn, the president of Curtiss-Wright, argued after the war that the production of numerous aircraft at a single factory "created great confusion on the part of the entire personnel." "In general, experience proved it most effective to 'type' a plant with one design, thus affording the opportunity for all concerned to concentrate on this one type and produce with the greatest efficiency and results." Curtiss-Wright, it should be noted, went from being one of the largest of the prewar manufacturers to, practically speaking, exiting the industry in the postwar era. See

Guy W. Vaughan, "Major Problems and Time Elements in Expanding the Aircraft Industry in World War II," in *Elements of American Air Power*, 89–96.

23. U.S. Army Air Forces, Air Technical Service Command, Logistics Planning Division, "Industrial Planning Project, Construction and Production Analysis, Douglas—Long Beach," [covering the A-20B, C-47, A-26, and C-74], March 1947, 68, 71, 77 (Air University Library, Maxwell AFB).

24. AAF Renton, exhibit 15. See also North American Aviation, "A Brief History of Operations," 49. A good description of one of Boeing's specialist groups, the propulsion unit, is found in: Discussion between Mr. G. Newton and Mr. R. Bradley, 11 October 1943, 8–9 (Model 345, B-29 Power Plant, 3214/5, Boeing Historical Archives).

25. U.S. Army Air Forces, Materiel Center, Wright Field, Production Engineering Section, Memorandum No. 43–1 (8 July 1942), "Block System of Production Releases," Committee Meeting Minutes, Vol. 2 (Model 345, Boeing Historical Archives).

26. AAF Renton, 38–39.

27. U.S. Army Air Forces, Air Materiel Command, Logistics Planning Division, Industrial Planning Section, "North American Aviation, Inc., Inglewood, California, Construction and Production Analysis," 63 (Air University Library, Maxwell AFB).

28. U.S. Army Air Forces, Air Technical Service Command, Logistics Planning Division, "Industrial Planning Project, Construction and Production Analysis, Boeing Wichita," June 1946, 45 (Air University Library, Maxwell AFB).

29. AAF Renton, 26.

30. Discussion between Mr. C. L. Egtvedt and Mr. A. Simonds, 14 October 1943, 9 (3214/5, Boeing Historical Archives).

31. D. J. Euler and E. T. Allen, "Progress Report on Current Research including Program," Report no. D-2974, 18 December 1940 (File 17, Box 2687, Boeing Historical Archives).

32. "Engineering Laboratory of the Bell Aircraft Corp." *Aero Digest* (November 1940): 124–26, 191.

33. Harold Mansfield, Boeing News Bureau, untitled press release on the Strato-Chamber, 25 June 1940; Boeing News Bureau, untitled press release on engine "go-round" house, 28 March 1942 (Subject file: Early Engineering, Research and Development through 1943, Boeing Historical Archives). The wind tunnel is described briefly in Euler and Allen, "Progress Report on Current Research including Program."

34. "The Allen Memorial Wind Tunnel," *Aero Digest* 45 (1 May 1944): 122–23, 134.

35. Boeing Company, "The Research Behind the Product," n.d. [ca. 1947–48] (Doc. no. 8415, File 13, Box 3154, Boeing Historical Archives).

36. "North American's Pressure Chamber," *Aero Digest* 51 (1 November 1945): 64–65, 134; "North American's New Wind Tunnel," *Aero Digest* 41 (August 1942).

37. F. P. Dillon, W. Doble, and H. A. Greenwald, "Lockheed Pressure Chamber: Testing Pressurizing Equipment for High Altitude Flight," *Aeronautical Engineering* 1 (April 1942): 13–15.

38. "Convair's New Wind Tunnel," *Aero Digest* 48 (15 January 1945): 110.

39. The strategy of establishing in-house engineering capabilities in order to interface with outside suppliers is noted in Giovanni Dosi, Mike Hobday, Luigi Marengo, and Andrea Prencipe, "The Economics of Systems Integration: Towards an

Evolutionary Interpretation," in *Business of Systems Integration*, ed. Andrea Prencipe, Andrew Davies, and Mike Hobday (New York: Oxford University Press, 2003), 95–113.

40. Boeing Company, "The Research Behind the Product."

41. See Johnson, *Secret of Apollo*, 74–79.

42. Martin J. Collins, *Cold War Laboratory: RAND, the Air Force, and the American State, 1945–1950* (Washington, DC: Smithsonian Institution Press, 2002); Johnson, *Secret of Apollo*.

9

Submarines Transformed: From Type XXI to Polaris, 1945–1960

Barton C. Hacker

Submarines proved themselves formidable weapon systems in both world wars. Until late in World War II, however, these so-called submarines were really diesel-powered boats optimized for surface speed and better termed "submersibles." Electric motors powered by storage batteries allowed them to hide underwater, but not for very long and at great cost in speed and maneuverability.[1] Even combat patrols during World War II routinely involved 90 percent (or more) surface operations, attacks included. In fact, wolf pack tactics, which required radio contact to coordinate the attack, could work only with surfaced boats.[2]

All that changed in the decade and a half after World War II. Radical innovations in submarine technology between 1945 and 1960 converted the submarine from a surface raider that could temporarily hide and sometimes fight underwater to a true underwater warship. Unprecedented improvements in submarine speed, maneuverability, and endurance forced every navy to rethink naval tactics. Existing techniques of antisubmarine warfare became obsolete overnight. Coping with the new threat quickly emerged as the special responsibility of attack submarines ("fast attacks," as the U.S. Navy calls them), while ballistic-missile submarines ("boomers" in U.S. Navy parlance) took on the new mission of deterring nuclear attack. How that came about is the chief focus of this chapter, with special attention to the crucial and surprising role of the United States Atomic Energy Commission's Livermore Laboratory in California.

SUBMARINES TRANSFORMED

The development of the modern submarine began with German Type XXI U-boats, a few of which became operational at the very end of World War II. Postwar American, British, and Soviet attack submarine models all adapted (rather than directly copied) Type XXI concepts of high underwater speed and prolonged underwater operation. Speed came from a new power source—a steam turbine—and a fish-shaped—rather than ship-shaped—hull optimized for underwater rather than surface performance. Endurance grew from a device that allowed a submerged submarine to breathe, the snorkel.[3]

Postwar U.S. submarines underwent far-reaching changes. Refurbished and streamlined hulls fitted with snorkels and improved power plants achieved dramatically better underwater performance.[4] These efforts culminated in the experimental USS *Albacore*, completed in 1953, capable of twenty-five knots surfaced and thirty-three knots submerged.[5] Nuclear power completed the revolution. The Type XXI had used heat from the catalytic breakdown of hydrogen peroxide to drive its steam turbine, but a decade later the United States had developed a nuclear reactor. Nuclear reactors are basically heat engines. In one sense, they merely succeeded combustion chambers as heat sources for steam turbines. Reactor heat drives the steam turbines that provide the submarine's actual power. The development of compact, highly reliable reactors for naval use in the early 1950s was a major technological achievement. Fuel capacity ceased to be a significant limit on underwater operation.[6]

Nuclear propulsion had developed in parallel to the revolution in submarine design. In fact, the navy had sponsored its first nuclear research as early as 1939, and an early postwar study concluded that nuclear propulsion would revolutionize naval warfare.[7] In 1946 the navy sent a small cadre of officers to spend a year at Oak Ridge, the nuclear research laboratory established in Tennessee by the Manhattan Project during World War II. There they learned the basics of nuclear reactor engineering. Among those sent to Oak Ridge was Capt. Hyman G. Rickover of the navy's Bureau of Ships. In 1948 Rickover became head of the newly established nuclear power branch within the Research Division of the Bureau of Ships. He had already demonstrated his engineering skills and bureaucratic acumen in wartime Washington. In short order, he would reveal himself to be a consummate politician as well.[8]

Rickover moved aggressively to develop the necessary power plant. Nuclear power was particularly attractive for submarines, but also presented some unique technical challenges. To the problems of materials, shielding, and radiation that all forms of nuclear propulsion faced, nuclear subs added the operation of a steam plant within an underwater shell, something the navy had never found practicable before. And all this before a single nuclear power reactor of any kind had ever gone into service. The Atomic Energy Act of 1946 created the Atomic Energy Commission as the civilian agency in charge of all nuclear development. Organizing the new agency and sorting out its priorities took time and work. When it finally established its Division of Reactor Development in early 1949, Rickover added the role of chief of its Naval Reactors Branch to his position in the Bureau of Ships. The navy and the Atomic Energy Commission agreed to assign the design of a prototype pressurized water reactor for submarines to the commission's Argonne Laboratory in Illinois, while contracting with Westinghouse for engineering development and production. Concurrent development of a prototype sodium-cooled reactor with Knolls Atomic Energy Laboratory in Ohio and General Electric began in 1950.[9]

In the early 1950s Rickover's organization directed the design and development of two land-based prototypes: (1) the Mark I version of the submarine thermal reactor, the pressurized water system; and (2) the Mark A version of the submarine intermediate reactor (SIR), the liquid-metal cooled system. Although both eventually powered submarines—the water-cooled Mark I in USS *Nautilus*, the sodium-cooled SIR in USS *Seawolf*—the pressurized water reactor was always considered the more promising and became in fact the model for all subsequent American nuclear-powered subs. The prototype first operated successfully in the summer of 1953 and *Nautilus* was under way on nuclear power in January 1955. She was the first true submarine, a vessel capable of swift underwater movement for weeks on end without ever coming near the surface.[10]

Although *Nautilus* and her immediate successors retained essentially conventional hulls, their nuclear power plants gave them speed and endurance so great as to make virtually all existing means of antisubmarine warfare obsolete. Their chief mission quickly became fighting other submarines. They also became immensely more efficient undersea warships in general.[11] The disparity grew still larger when nuclear power plants were installed in *Albacore*-style hulls optimized for submerged speed. The first was *Skipjack*, laid down by Electric Boat even before the company had completed *Nautilus*. Commissioned in 1959, it defined the basic shape of all subsequent U.S. nuclear submarines.[12] Although fast attack submarines like *Skipjack* and its successors remained, in essence, more capable performers of traditional submarine roles, the new technology they employed also made possible a very different kind of submarine, an undersea launching platform for nuclear-armed missiles.

LIVERMORE AND NOBSKA

In September 1952 the newly formed Livermore branch of the University of California Radiation Laboratory officially opened its doors. It had a contract with the United States Atomic Energy Commission (AEC), but little else was defined. The exact nature of the branch's mission remained open—neither university nor commission concerned itself overmuch with a formal mission statement—though the broad outlines seemed clear enough: Livermore would address aspects of designing and testing thermonuclear weapons, as well as certain related basic research. Initially, at least, the new laboratory filled a supporting role. Despite innumerable references to Livermore as the nation's second nuclear weapons research laboratory, it could as yet only aspire to a status that Los Alamos, its predecessor and prospective rival, had long since achieved. Los Alamos was the laboratory established under Manhattan Project auspices in World War II to design and build the first atomic bombs.[13]

Lacking either a precisely defined mission or any great expertise in nuclear weapons design, Livermore began modestly. Novice designers tried to rework ideas Los Alamos had bypassed, to explore areas Los Alamos had, for whatever reasons, left untouched. Herbert York, the laboratory's first director, set the "working philosophy . . . at the very beginning" that his colleagues "readily accepted." It

> called for always pushing at the technological extremes. We did not wait for higher government or military authorities to tell us what they wanted and only then seek to supply it. Instead, we set out from the start to construct nuclear explosive devices that had the smallest diameter, the lightest weight, the least investment in rare materials, or the highest yield-to-weight ratio or that otherwise carried the state of the art beyond the currently explored frontiers. We were completely confident that the military would find a use for our product after we proved it, and that did indeed usually turn out to be the case.[14]

Enthusiasm notwithstanding, the laboratory's early track record looked none too promising.

Livermore-designed devices tested during 1953 in Nevada and 1954 at Bikini had yields so far below expectation as to prompt some jeering observers to call them "fizzles." But the young scientists and engineers learned from their mistakes and soon turned things around. The breakthrough for fission designs came in 1955 at the Nevada Test Site in Operation Teapot, for thermonuclear designs in 1956 at the Pacific Proving Ground in Operation Redwing. One of Livermore's successful Redwing tests was a small H-bomb design.[15] That test in July 1956, came at very nearly the same time that Edward Teller, one of the laboratory's founders and now its associate director for physics, made an electrifying announcement half a world away.

The scene was Nobska Point at Woods Hole, Massachusetts, where the National Academy of Sciences' Committee on Undersea Warfare had organized a summer study meeting. Such meetings had become, since the late 1940s, a common means for military organizations to enlist academic experts in thinking about large problems.[16] The 1956 study, code-named Project Nobska after its meeting place, was sponsored by the Office of Naval Research to ponder the future of undersea warfare. Sixty scientists and engineers—military, industrial, and academic—met from mid-June to mid-September. Following a two-week briefing from navy experts at the outset, they divided into fourteen groups to study specific areas, identify the main problems, and suggest appropriate research and development to solve them.[17]

The panel on antisubmarine weapons included Livermore's Edward Teller. He proposed arming torpedoes with nuclear warheads, surprising his listeners by claiming that high yields could soon be expected from designs scaled to torpedo size and weight. Dismissing the bulk of current high-yield nuclear weapons, he invoked the technological trend toward major weight reductions in each warhead generation; if lighter warheads might not yet be available, they soon would be.[18] Under questioning, Teller also suggested the relevance of this trend for the navy's planned Fleet Ballistic Missile. "Why use a 1958 warhead in a 1965 weapon system?" he asked rhetorically.[19] The final report of Project Nobska strongly endorsed this view, judging it

> reasonable to expect that warhead weights, for a given yield, will be decreased significantly during the next seven years. Missile designers should ask for new warheads to meet their design requirements instead of designing new missiles around existing warheads, which would be obsolete by the time of missile production.[20]

Teller's forecast proved to be just what the officers in the navy's new Special Projects Office wanted to hear.

THE SPECIAL PROJECTS OFFICE AND THE MISSILE RACE

Established in late 1955 to solve the problems of launching intermediate-range ballistic missiles (IRBMs) from ships, the navy's Special Projects Office owed its existence to Arleigh Burke, the young admiral who had leapfrogged dozens of senior officers to become chief of naval operations in August 1955. He strongly believed that sea-launched missiles were both technically feasible and highly desirable. Their successful development would contribute mightily to American security as a nearly invulnerable nuclear deterrent. Burke handpicked the new office's director, Rear Adm. William F. Raburn, and backed him to the hilt.[21] As a latecomer to the missile business, however, the navy found itself in an awkward position.

In February 1955, after an intensive five-month study of current military technology and the Soviet nuclear threat to the United States, President Dwight D. Eisenhower's Technological Capabilities Panel, chaired by MIT president James E. Killian Jr., submitted its report to the National Security Council. It urged the highest national priority for air force development of an intercontinental ballistic missile (ICBM). The report also favored work on intermediate-range missiles, both land and sea based, largely because such major problems as guidance and reentry might be more easily solved for shorter ranges. Several months later the National Security Council concurred, and the president agreed.[22]

Until Burke's appointment as chief of naval operations, however, the navy had not evinced much interest in sea-based missiles, leaving the field of long-range missiles to the army and air force.[23] Many naval officers were, in fact, hostile or indifferent to such applications, preferring to see navy work focused on missiles suited to specific navy needs. Strategic deterrence was not perceived as a navy job.[24] By late 1955 the air force not only had control over development of intercontinental missiles, but was on its way to capturing land-based intermediate-range ballistic missiles as well; it had three of the four missile projects the Department of Defense was willing to approve: the Atlas ICBM, its backup Titan, and the Thor IRBM.[25]

The army still clung to its IRBM #2 (formally named Jupiter in March 1956) as Thor's backup, but the recommended slot for a sea-launched IRBM remained unfilled. Because the Department of Defense would not, for reasons of economy, approve a fifth missile project, the navy became the army's reluctant partner in the Jupiter program. Although relations between the two services seemed cordial enough, the liquid-fueled Jupiter was not a promising candidate for launching at sea. To loft its three-thousand-pound (1,400 kg) payload fifteen hundred miles (2,400 km) required, even in a proposed solid-fuel navy version called Jupiter S, an eighty-ton missile forty-four feet (14 m) long and ten feet (3 m) in diameter.[26]

Special Projects, in short, stood ready to welcome just what Teller promised and Project Nobska endorsed. Extrapolating major missile subsystem technologies, the Nobska panel on the Strategic Uses of the Undersea could find no decisive advantage in any single advance, but saw that collectively they promised an extremely effective weapon system by the early 1960s. It recommended the navy build a solid-fuel missile of eight to fifteen tons with a range of one thousand to fifteen hundred miles (1,600 to 2,400 km) carrying a fractional-megaton warhead. Opposition from other summer study participants failed to squelch the panel's recommendation, but it appeared in the final study report as only one among many others. It was nonetheless, as events proved, the only one that mattered.[27] It mattered because it promptly found a constituency. If payloads one-third the weight of existing packages could reasonably be expected within a few years to provide significant nuclear yields, then navy hopes for a sea-based deterrent brightened greatly.

Relatively low-yield, lightweight thermonuclear warheads had become one focus of work at Livermore as early as 1953. During the mid-1950s, all three services evinced interest in the potential of warheads with yields between 250 kilotons and 2 megatons that were light enough to be carried by fighter-bombers, unmanned aircraft, or missiles. Before 1956, however, no formal military requirement existed, and Livermore's efforts remained theoretical and experimental.[28] Jack Rosengren, a key figure in Livermore's Polaris project, later described the technological problem as it appeared in the mid-1950s. "High-yield warheads were simply too large and heavy to be thrown any real distance by any U.S. missiles that would fit on submarines," he recalled. "It was expected that it would take years to develop even a large, liquid-fueled missile that could carry a megaton-class warhead the 1,500 miles that the planners considered necessary."[29]

Moving cautiously, the navy cross-checked Teller's version of technical trends in warhead design with the weapons laboratories, the Atomic Energy Commission, and private contractors. Although Los Alamos experts judged Livermore's projections overly optimistic, everyone seemed to agree about the general trend toward smaller and lighter nuclear weapons.[30] The Special Projects Office found enough to warrant action, and the Polaris missile program was fairly launched.[31]

In October 1956 key navy, business, and academic officials met "to develop the plans and identify the tradeoffs required to select the envelope parameters necessary to enable concurrent development of the submarine and the weapon system," as R. A. Fuhrman described it. "Each individual was authorized to commit his parent organization without lengthy referral to higher management."[32] Fuhrman joined Lockheed Missiles & Space Company in 1959 to manage Polaris systems integration and later became the company's president. Although Lockheed had long been a major defense contractor, it established its missile division only in 1954. Like the Livermore laboratory and the Special Projects Office itself, Lockheed was another newcomer to the strategic missile business, and Polaris was its first prime contract.[33]

POLARIS WARHEAD DEVELOPMENT

By the end of 1956 the secretary of defense had authorized the navy to withdraw from the Jupiter program and to devote itself to developing Polaris. Until then, Raburn's Special Projects Office had been responsible only for ship-based aspects of the weapon system; now it took charge of the entire program, including the nuclear-powered submarines that would carry the new missiles.[34] On 17 June 1957, Admiral Burke formally approved the characteristics planned for the Polaris submarine, which received the highest priority of any current navy project.[35]

Preliminary design studies of the Polaris missile, from which the final version did not greatly differ, defined a fifteen-ton solid-fuel missile 28.5 feet (8.77 m) long and 4.5 feet (1.4 m) in diameter, with a range of fifteen hundred nautical miles (2,400 km) for a nine-hundred-pound (410 kg) payload. During the first half of 1957, Livermore completed a feasibility study of the required warhead. Developmental engineering was the next step. Officially, Livermore received the assignment in July 1957.[36]

Moving the submarine-launched ballistic missile system from plausible speculation to feasible technology required advances across a broad front ranging from nuclear propulsion to inertial guidance. A viable lightweight warhead was nonetheless a crucial technical factor, though still in 1956 only a persuasive promise. But it was a promise from Edward Teller, backed by other Livermore scientists and the Atomic Energy Commission, that lent it political potency. As political scientist Harvey Sapolsky, author of one of the earliest and still the best published study of Polaris development, observes, the alleged "1956 warhead breakthrough" became a staple of early Special Projects Office literature, the dramatic change "that allowed the FBM system to switch from the large Jupiter missile to the smaller Polaris missile."[37]

For the still struggling Livermore branch of the Berkeley Radiation Laboratory, the Polaris warhead became the first substantial weapon assignment. Because Livermore needed the job and the navy needed the technology, it was, as one Polaris warhead designer at Livermore later termed it, "love at first sight."[38] Livermore director York deemed the navy project a vindication of his development philosophy. What became the "Polaris warhead started out simply as the lightest-weight thermonuclear warhead we could figure out how to build. After work on it was well under way, the Navy requirement for a lightweight warhead for its first submarine-launched missile, the Polaris, came along. They were a perfect match."[39]

Livermore had specific warhead design ideas ready for testing in Operation Plumbbob, the summer 1957 test series in Nevada.[40] Earlier work on small, low-yield fission weapons for the army had not only produced a successful artillery warhead design but also spurred Livermore to develop more advanced computer codes for modeling physical phenomena in nuclear explosions. Both the computer codes and the earlier experience proved indispensable to the laboratory's quick start on Polaris. Livermore's Plumbbob tests resulted in the Robin family of primaries, which were especially well suited to the volumetric constraints of strategic missiles. Thermonuclear weapons require at least two stages: a primary fission device to ignite the thermonuclear reaction in the secondary. Polaris used a Robin primary, as did many other later missiles.[41]

The Polaris program suddenly became far more urgent after 4 October 1957, when the first *Sputnik* went into orbit. Many took that beeping satellite as evidence that the Soviet Union already had what amounted to a working intercontinental missile, a belief reinforced by the much larger payload launched a month later.[42] Polaris development shifted to higher gear. Acting on a perceived emergency, the navy now intended to have its first missile-armed submarines at sea by 1960, five years earlier than originally planned, though with a missile of only twelve-hundred-mile (1,900 km) range.[43]

To accelerate production, the Bureau of Ships decided to use the barely completed hull of *Scorpion*, a *Skipjack*-class submarine under construction at Electric Boat. The builders installed a 140-foot (43 m) missile section with sixteen launch tubes between *Scorpion*'s bow and stern, and the ship was renamed *George Washington*.[44] The key to achieving the goal of a Polaris payload 30 percent of Jupiter's three thousand pounds (1,400 kg) lay in integrating Livermore's warhead with Lockheed's reentry body. This was largely the work of the Atomic Energy Commission's third weapons laboratory, Sandia in New Mexico, which specialized in turning experimental results into engineering practice. Choices about materials and design required thinking about such dual functions as structural integrity and thermal protection at the same time, "a then-unique and unconventional approach," according to Lockheed's Fuhrman.[45]

With the design of the warhead primary settled, Operation Hardtack—Phase 1 in the Pacific during the spring and summer of 1958, Phase 2 in Nevada during early fall—provided Livermore the opportunity to test its design for the Polaris secondary.[46] The name of this new design remains classified, but it made novel use of oralloy—an enriched uranium alloy from Oak Ridge—to obtain significant improvements in yield-to-weight and yield-to-volume ratios. These characteristics made secondaries of this type especially attractive for later use in multiple reentry vehicles (MRVs) and then multiple independently targetable reentry vehicles (MIRVs).[47]

A safety test in Hardtack Phase 2 revealed a problem with the Polaris primary. High explosives surround the primary's fission core. Simultaneously detonated, these explosives implode the fissile material to produce an explosive nuclear reaction. One-point safety requires that neither by accident or intent can the nuclear reaction be initiated by setting off the high explosives at a single point, as, for example, might occur if a bullet struck the weapon. Unfortunately, the warhead's primary was not one-point safe. The mid-October Hardtack safety test, code-named Neptune, produced a yield of 115 tons and demanded a fix.[48]

Compounding the problem was the immanent onset of the nuclear test moratorium at the end of the month. Although Livermore devised a mechanical safing system and had it ready for testing in two weeks, the test was scheduled for the last day before the moratorium. Atmospheric conditions delayed the test, which was finally cancelled.[49] The safety question could not be fully resolved, but computer analyses confirmed by small, no-yield hydrodynamic tests satisfied the weaponeers and the customer. In July 1960 the navy accepted delivery of the first sixteen weapons. Commissioned at the end of 1959, USS *George Washington*, the first Polaris submarine, was ready to receive them. Four months later she sailed on her maiden patrol with sixteen armed missiles in their launch tubes.[50]

When the moratorium ended, the Polaris missile system proved itself in an operational test called Frigate Bird, part of Operation Dominic in the Pacific. On 6 May 1962 the submerged USS *Ethan Allen* launched a stockpile Polaris missile with a live nuclear warhead over the Pacific toward Christmas Island, seventeen hundred miles (2,700 km) away. Twelve and a half minutes and just under twelve hundred miles (2,000 km) later, the warhead exploded as programmed between ten thousand and fifteen thousand feet (3,000 and 4,600 m) high.[51] Commissioned in November 1960, *Ethan Allen* was the first Polaris submarine designed from the keel up for the job. Thirty miles (48 km) from the airburst at periscope depth, the waiting USS *Carbonero*, a conventional submarine, captured the towering mushroom cloud on film.[52]

Frigate Bird remains the only full-scale operational test from launch through detonation ever conducted for a U.S. strategic nuclear missile. Designated W-47, the Livermore-designed Polaris warhead was hardly flawless, which led to a series of modifications. The problems of mechanical safing, in particular, persisted. What could not be fixed could be lived with, and it remained in service until 1974, finally retired only in favor of an improved Livermore design.[53]

Polaris lay very near the heart of the navy's effort to recast its role for the last half of the twentieth century. The early deployment of the Polaris missile system allowed the navy to demonstrate the feasibility of a sea-based missile deterrent before the air force preempted the field. In winning its race against the air force, the navy could rely on a long-standing relationship with Electric Boat to provide the submarine, but had to turn to two novices—Lockheed and Livermore—for the missile and its warhead. Both performed brilliantly. Although only a small part of the overall program, Livermore's work on the missile warhead was absolutely vital. It was also a triumph. Extraordinary success in meeting budgets and schedules, as well as specifications, in its first full-fledged weapon program marked a major turning point in laboratory history.[54]

Yet nothing was foreordained. In the mid-1950s Lockheed, Livermore, and the navy were all newcomers to the competitive world of strategic missiles. That all three also challenged the limits of existing technology in their programs made them natural allies, and the partnership flourished because institutional imperatives matched attainable technology. When Polaris proved a model R&D program in every significant respect, all partners benefited enormously. But the fledgling laboratory had far more to gain than the well-established company or the senior service, and so reaped the greater rewards. Polaris has regularly been described as Livermore's coming of age, paving the way for the further assignments that made it the nation's premier designer of strategic warheads for years to come.

The underwater endurance and stealth of nuclear-powered submarines, which could lurk undetected beneath the waves for weeks on end, promised a virtually invulnerable missile launch platform. Uniting the endurance and stealth of nuclear power with the new technology of long-range guided missiles, the Polaris fleet ballistic missile system created for the United States a highly credible nuclear deterrent against nuclear attack. Both submarines and missiles moved from idea to reality in two of the most successful research and development programs in recent history. Those successes in the midst of the Cold War completed the transformation of the submarine that had begun in the closing months of World War II.

NOTES

1. James L. McVoy, Virgil W. Rinehart, and Prescott Palmer, "The Roosevelt Resurgence (1933–1941)," in *Naval Engineering and American Seapower*, ed. Randolph W. King (Baltimore, MD: Nautical & Aviation Publishing Company of America, 1989), 171–77; Norman Friedman, *Submarine Design and Development* (Annapolis, MD: Naval Institute Press, 1984), chap. 3, "The Era of the Submersible, 1900–1945."

2. Edward M. MacCutcheon, "World War II Development and Expansion, 1942–1945," in *Naval Engineering and American Seapower*, 235–39; John Terraine, *The U-Boat Wars, 1916–1945* (New York: G. P. Putnam's Sons, 1989); V. E. Tarrant, *The U-Boat Offensive, 1914–1945* (Annapolis, MD: Naval Institute Press, 1989).

3. Siegfried Breyer, *German U-Boat Type XXI*, trans. Ed Force (Atglen, PA: Schiffer Military/Aviation History, 1999). On the "big brother" of Type XXI, see Eberhard Rössler, *U-Boottyp XXIII* (2d ed.; Bonn: Bernard & Graefe Verlag, 2002). For the whole range of innovative German U-boat designs, see Eberhard Rössler, *The U-Boat: The Evolution and Technical History of German Submarines*, trans. Harold Erenberg (Annapolis, MD: Naval Institute Press, 1981), chap. 10, "U-Boat Development at the End of the War."

4. Norman Friedman, *U.S. Submarines since 1945: An Illustrated Design History* (Annapolis, MD: Naval Institute Press, 1994), chap. 3, "The Fast Submarine"; Gary E. Weir, *Forged in War: The Naval-Industrial Complex and American Submarine Construction, 1940–1961* (Washington, DC: Naval Historical Center, 1993), chap. 5, "Cold War Challenges, 1946–1951"; Norman Polmar, *The American Submarine* (Cambridge, MA: Patrick Stephens, 1981), chap. 6, "Guppies and Other Fishes."

5. Weir, *Forged in War*, chap. 6, "From Hartwell to Albacore, 1946–1953"; David Miller, *Submarines of the World* (New York: Orion Books, 1991), "Albacore Class," 38–39; Robert P. Largess and James L. Mandelblatt, *U.S.S. Albacore: Forerunner of the Future*, Portsmouth Marine Society Publication 25 (Portsmouth, NH: P. E. Randall, 1999).

6. Friedman, *Submarine Design and Development*, chap. 9, "Propulsion—Conventional and Nuclear."

7. Joseph-James Ahern, "'We Had the Hose Turned on Us!' Ross Gunn and the Naval Research Laboratory's Early Research into Nuclear Propulsion, 1939–1946," *Historical Studies in the Physical and Biological Sciences* 33, no. 2 (2003): 217–36.

8. On Rickover, see especially Francis Duncan, *Rickover and the Nuclear Navy: The Discipline of Technology* (Annapolis, MD: Naval Institute Press, 1990); Theodore Rockwell, *The Rickover Effect: The Inside Story of How Adm. Hyman Rickover Built the Nuclear Navy* (1992; reprint, New York: John Wiley & Sons, 1995).

9. Richard G. Hewlett and Francis Duncan, *Nuclear Navy, 1946–1962* (Chicago: University of Chicago Press, 1974), chaps. 3–6; Friedman, *U.S. Submarines since 1945*, chap. 6, "The Nuclear Revolution"; Willis C. Barnes, "Korea and Vietnam (1950–1972), in *Naval Engineering and American Seapower*, 278–81. See also John W. Crawford and Steven L. Krahn, "The Naval Nuclear Propulsion Program: A Brief Case Study in Institutional Constancy," *Public Administration Review* 58 (1998): 159–82.

10. Jeffrey L. Rodengen, *The Legend of Electric Boat* (Ft. Lauderdale, FL: Write Stuff Syndicate, 1994), chap. 8, "The Cold War and the *Nautilus*"; Weir, *Forged in War*, chap. 8, "Under Way on Nuclear Power, 1953–1956"; Miller, "Nautilus," in *Submarines of the World*, 40–41.

11. Robert A. Kuenne, *The Attack Submarine: A Study in Strategy* (New Haven, CT, and London: Yale University Press, 1965), chap. 9, "The Nuclear Attack Submarine"; Friedman, *U.S. Submarines since 1945*, chap. 4, "The ASW Submarine."

12. Hewlett and Ducan, *Nuclear Navy*, 268–70, 307; Friedman, *U.S. Submarines since 1945*, 130, 234.

13. Ruth R. Harris and Richard G. Hewlett, "The Lawrence Livermore National Laboratory: The Evolution of Its Mission, 1952–1988" (Rockville, MD: History Associates, 30 May 1990), 3–5. For Livermore's origins, see Barton C. Hacker, "A Modest Proposal: Berkeley and the Second Weapons Laboratory, 1949–1954," presented at the Symposium on Postwar Big Science and Technology, XIXth International Congress of the History of Science, Zaragoza, Spain, 22–29 August 1993. On the wartime role of Los Alamos, see Lillian Hoddeson et al., *Critical Assembly: A Technical History of Los Alamos During the Oppenheimer Years, 1943–1945* (New York: Cambridge University Press, 1993).

14. Herbert F. York, *Making Weapons, Talking Peace: A Physicist's Odyssey from Hiroshima to Geneva* (New York: Basic Books, 1987), 75.

15. For accounts of the relevant tests, see Chuck Hansen, *US Nuclear Weapons: The Secret History* (Arlington, TX: Aerofax, 1988), 39, 66–68, 72–74.

16. For an overview of navy-sponsored summer studies, including Project Nobska, see J. R. Marvin and F. J. Weyl, "The Summer Study," *Naval Research Review* 20 (August 1966): 1–7, 24–28. See also David C. Elliot, "Project Vista and Nuclear Weapons Europe," *International Security* 11 (1986): 163–83; and Ann Finkbeiner, *The Jasons: The Secret History of Science's Postwar Elite* (New York: Viking, 2006).

17. Gary E. Weir, "The Next Time You Go for a Walk . . . Project Nobska, 1956," *Pull Together: Newsletter of the Naval Historical Foundation and the Naval Historical Center* 33, no. 1 (Spring/Summer 1994): 5–8.

18. Harvey M. Sapolsky, *The Polaris System Development: Bureaucratic and Programmatic Success in Government* (Cambridge, MA: Harvard University Press, 1972), 28–29; Wyndham D. Miles, "The Polaris," *Technology and Culture* 4 (1963): 481.

19. Reported by William F. Whitmore, "Military Operations Research—A Personal Retrospect," *Operations Research* 9 (March–April 1961): 263, as quoted in Sapolsky, *Polaris System Development*, 30. See also Miles, "The Polaris," 481.

20. "Project Nobska: The Implications of Advanced Design on Undersea Warfare," Final Report, 2 vols., 1 December 1956–1 March 1957, 2:260, as quoted in Weir, *Forged in War*, 248. The Project Nobska report remains classified (Weir, *Forged in War*, 211n).

21. David Alan Rosenberg, "Arleigh Albert Burke, 17 August 1955–1 August 1961," in *The Chiefs of Naval Operations*, ed. Robert William Love Jr. (Annapolis, MD: Naval Institute Press, 1980), 277–79; Hewlett and Duncan, *Nuclear Navy*, 266; Vincent Davis, *The Politics of Innovation: Patterns in Navy Cases*, Social Science Foundation and Graduate School of International Studies, University of Denver, Monograph Series in World Affairs, vol. 4, no. 3 (Denver, CO, 1967), 26–27.

22. James R. Killian Jr., *Sputnik, Scientists, and Eisenhower: A Memoir of the First Special Assistant to the President for Science and Technology* (Cambridge, MA, and London: MIT Press, 1977; paperback ed., 1982), 70, 76; Davis, *Politics of Innovation*, 23–24. On the role of the NSC, see Stanley L. Falk, "The National Security Council of Truman, Eisenhower, and Kennedy," *Political Science Quarterly* 74 (1964): 403–34; Falk, *The National Security Structure* (Washington, DC: Industrial College of the Armed Forces, 1967), chap. 3, "The National Security Council."

23. Davis, *Politics of Innovation*, 21–23.

24. Hewlett and Duncan, *Nuclear Navy*, 223; Rosenberg, "Arleigh Albert Burke," 277–78; Davis, *Politics of Innovation*, 24–26.

25. Jacob Neufeld, *The Development of Ballistic Missiles in the United States Air Force, 1945–1960* (Washington, DC: Office of Air Force History, 1989); Edmund Beard, *Developing the ICBM: A Study in Bureaucratic Politics* (New York: Columbia University Press, 1976); G. Harry Stine, *ICBM: The Making of the Weapon That Changed the World* (New York: Orion Books, 1991).

26. Michael H. Armacost, *The Politics of Weapons Innovation: The Thor-Jupiter Controversy* (New York and London: Columbia University Press, 1969), 108–10.

27. So concluded the National Academy of Science's Committee on Undersea Warfare when it reviewed Project Nobska a year later; see Committee on Undersea Warfare, *Review of Project Nobska, Final Report, August 5–9, 1957* (Washington, DC: National Academy of Sciences, 1 November 1957), as discussed in Sapolsky, *Polaris System Development*, 30–31.

28. Sybil Francis, "Warhead Politics: Livermore and the Competitive System of Nuclear Weapons Design" (PhD diss., Massachusetts Institute of Technology, 1995), 118–19. Based on extensive research in archives and interviews with leading weaponeers at Livermore, Los Alamos, and elsewhere in the weapons complex, this dissertation has also been issued under the same title as a Livermore report: UCRL-LR-124754, June 1995.

29. As quoted in "Polaris: Livermore Comes of Age," *Newsline* [Lawrence Livermore National Laboratory] (September 1972): 24–25, at 24.

30. Herbert York (Livermore Director) to Alfred D. Starbird (Director of Military Application, Atomic Energy Commission), 5 November 1956, with marginal notes by Los Alamos experts; Norris E. Bradbury to Willard F. Libby, 15 June 1959, as quoted in Francis, "Warhead Politics," 118–19.

31. Strategic Systems Program Office, *FBM* [Fleet Ballistic Missile] *Chronology: FBM Facts/Chronology: Polaris, Poseidon, Trident* (Washington, DC: Navy Department, 1986), 24; Sapolsky, *Polaris System Development*, 31–32.

32. R. A. Fuhrman, "Fleet Ballistic Missile System: Polaris to Trident," von Kármán Lecture for 1978, American Institute of Aeronautics and Astronautics annual meeting, Washington, DC, February 1978, 4. This lecture was subsequently published in *Journal of Spacecraft* 15 (September–October 1978): 265–86.

33. Gordon Adams, *The Iron Triangle: The Politics of Defense Contracting* (New York: Council on Economic Priorities, 1981), 317; Kenneth A. Bertsch and Linda S. Shaw, *The Nuclear Weapons Industry* (Washington, DC: Investor Responsibility Research Center, 1984), 206–209.

34. U.S. Cong., House, Comm. on Government Operations, *Organization and Management of Missile Programs: Eleventh Report*, House Rpt. No. 1121, 86th Cong., 1st Sess., 1959, 36; Strategic Systems Program Office, *FBM Chronology*, 25.

35. Hewlett and Duncan, *Nuclear Navy*, 310.

36. Francis, "Warhead Politics," 120–21; Fuhrman, "Fleet Ballistic Missile System," 11; Strategic Systems Program Office, *FBM Chronology*, 25.

37. Sapolsky, *Polaris System Development*, 47. On missile guidance as related to Polaris, see Donald Mackenzie, *Inventing Accuracy: A Historical Sociology of Missile Guidance* (Cambridge, MA, and London: MIT Press, 1990), 134–52.

38. Sybil Francis interview with Jack W. Rosengren, Livermore, 4 March 1991, as quoted in Francis, "Warhead Politics," 120.

39. York, *Making Weapons*, 76.

40. Frank H. Shelton, *Reflections of a Nuclear Weaponeer* (Colorado Springs, CO: Shelton Enterprise, 1988), chap. 8, 21 (Table 8-3: "Operation Plumbbob—Summer 1957, Nevada Test Site").

41. Francis, "Warhead Politics," 127.

42. Hewlett and Duncan, *Nuclear Navy*, 310–14. For recent assessment of *Sputnik*'s impact and the missile gap controversy, see Robert A. Divine, *The Sputnik Challenge* (New York: Oxford University Press, 1993); McGeorge Bundy, *Danger and Survival: Choices about the Bomb in the First Fifty Years* (New York: Random House, 1988), 331–51.

43. Fuhrman, "Fleet Ballistic Missile System," 5; Strategic Systems Program Office, *FBM Chronology*, 25.

44. Rodengen, *Legend of Electric Boat*, 116–17; Hewlett and Duncan, *Nuclear Navy*, 315; Weir, *Forged in War*, 249–51; Friedman, *U.S. Submarines since 1945*, 199.

45. Fuhrman, "Fleet Ballistic Missile System," 7. On Sandia and its work, see Necah Steward Furman, *Sandia National Laboratories: The First Decade* (Albuquerque: University of New Mexico Press, 1990); Leland Johnson, *Sandia National Laboratories: A History of Exceptional Service in the National Interest* (Albuquerque, NM: Sandia National Laboratories, 1997).

46. Shelton, *Reflections of a Nuclear Weaponeer*, chap. 9, 19 (Table 9-1: "Operation Hardtack Phase I—1958"), 45 (Table 9-3, "Operation Hardtack Phase II—1958"); Hansen, *US Nuclear Weapons*, 75–81.

47. Francis, "Warhead Politics," 125–26. See also Ted Greenwood, *Making the MIRV: A Study of Defense Decision Making* (Lanham, MD: University Press of America, 1988).

48. Hansen, *US Nuclear Weapons*, 204.

49. Barton C. Hacker, *Elements of Controversy: The Atomic Energy Commission and Radiation Safety in Nuclear Weapons Testing, 1947–1974* (Berkeley and Los Angeles: University of California Press, 1994), 196–200.

50. Hewlett and Duncan, *Nuclear Navy*, 315; Friedman, *U.S. Submarines since 1945*, 235; Strategic Systems Program Office, *FBM Chronology*, 27.

51. Austin W. Betts (AEC Director of Military Application), Dominic Test Bulletins no. 19, "Frigate Bird Event," 7 May 1962, and no. 27, " . . . Frigate Bird Data," 28 May 1962; Allen H. Anderson to Glenn T. Seaborg (AEC Chairman), "Frigate Bird Test Results," 23 May 1962; Seaborg to McGeorge Bundy, 15 April 1963, with attached "Summary of Results of the 1962 Atmospheric Nuclear Test Series."

52. Hansen, *US Nuclear Weapons*, 85; Hewlett and Duncan, *Nuclear Navy*, 317; Friedman, *U.S. Submarines since 1945*, 230, 235.

53. Hansen, *US Nuclear Weapons*, 203–205. See also Graham Spinardi, *From Polaris to Trident: The Development of US Fleet Ballistic Missile Technology* (Cambridge: Cambridge University Press, 1994), 58–63.

54. "Polaris: Livermore Comes of Age," special issue entitled "20 Years in Livermore," *Newsline* [LLNL] 3, no. 9 (September 1972): 24–25.

10

The World According to GARP: Scientific Internationalism and the Construction of Global Meteorology, 1961–1980

Erik M. Conway

During the late 1960s and throughout the 1970s, a series of large-scale international meteorological field expeditions took place. These field experiments, a product of Cold War scientific internationalism, were intended to examine meteorological phenomena with the help of a new scientific tool: orbital satellites. Called the Global Atmospheric Research Program, at its height in the mid-1970s it involved contributions from seventy nations.

GARP's purpose was the development of global weather prediction. It sponsored large-scale field experiments designed to provide the understanding needed to better model tropical circulation, energetics, and the Indian monsoon. The program's founder and principal backer was the United States, whose scientists sought global data for their model research, whose politicians found it useful to accede to their demands, and whose military sought global prediction for its own use—although it remained conspicuously uninvolved in GARP. GARP's American leaders were thus able to serve their scientific ideals, their profession, and their nation simultaneously.

On 1 April 1960, the United States launched the world's first weather satellite, TIROS I. With the space age scarcely two years old, TIROS was a highly public success for the American space program, which had been troubled by failures in the face of Soviet successes. The new American space agency, the National Aeronautics and Space Administration (NASA), needed such a victory.[1] TIROS, fortunately, was a particularly powerful one. The weather was something everyone experienced, and the "cloud pictures" radioed back to Earth had an immediate relevance to people that the other early scientific results of the space age did not.

To meteorologists, TIROS offered the power of a "global view." Weather was a phenomenon that did not respect the arbitrary political borders humans had created to order their world. That fact had motivated European and American meteorologists to make occasional efforts to construct international weather reporting networks; after World War II, they had finally succeeded in constructing international networks of relative permanency.[2] Yet these only covered the landmasses of certain Northern Hemisphere nations and a few weather ships that they supported, leaving the majority of Earth unobserved. Better weather forecasting required more information, particularly from the oceanic atmosphere, and TIROS, and other satellites like it, could provide it.

TIROS also represented an opportunity for the American government to showcase its postwar commitment to an internationalist foreign policy. Clark Miller recently argued that the internationalist bent of meteorological scientists must be placed within the larger context of American postwar politics.[3] In a substantial departure from pre–World War II policies, postwar American political leaders generally supported the construction of international institutions designed to promote liberal, democratic, capitalist values in the world. These leaders, from both major political parties, believed that networks of treaties and institutions would lead to a more stable world and thus enhance American security. Miller contends that the international meteorological networks that evolved during the first postwar decades were part of that internationalist program.

The American government therefore decided to enmesh TIROS in an international framework, leaving the satellites under American control but making their data available internationally. It also decided to sponsor an international research program around the satellites, to bring about a truly global meteorology based upon satellites and numerical weather prediction. This became known as the Global Atmospheric Research Program—GARP. During this period, the United States military sought the ability to fight wars globally.[4] It had established a weather-reporting network during World War II that had been almost global, giving its leaders access to forecasts of a quality never achieved before.[5] The high cost of this effort brought it to an end during the immediate postwar years, but the development of numerical weather prediction under a joint military and Weather Bureau project during the 1950s, and the nearly simultaneous development of satellite technology, promised to restore a global forecasting ability at a substantially lower cost.[6]

This new capability needed to be checked against more traditional data and forecasting methods, however, requiring the United States to seek partners. GARP served this purpose, too, providing the United States access to other nations' data while also defraying some of the costs. While GARP's American sponsors were the Weather Bureau, the National Science Foundation, and NASA, not the Defense Department, the American military stood to gain substantially anyway, as the results became available to its own researchers and operational forecasters. GARP thus served to advance meteorology, American foreign policy, and military capabilities simultaneously.

THE WEATHER FROM SPACE

TIROS stood for Television-InfraRed Observation Satellite, a moniker that accurately reflected its origin in a surveillance satellite program. In 1956 the U.S. Air Force had initiated a program to develop a reconnaissance satellite. Lockheed won the air force contract, leading RCA, the unsuccessful bidder, to seek army funding. The army accepted RCA's proposal and initiated JANUS, a reconnaissance satellite program of its own.[7] When the Advanced Research Projects Agency was formed in 1958, JANUS became an ARPA project. At the same time, responsibility for reconnaissance satellites was removed from the army and JANUS had its television cameras "detuned" to a resolution that would serve for meteorology but not for intelligence gathering.[8] Its name was also changed to TIROS.

The day before signing NASA's founding legislation, President Dwight D. Eisenhower decided that meteorological satellite development should be a civilian endeavor. It was transferred to NASA on 13 April 1959. NASA leaders also sought to bring researchers with relevant expertise into the agency, recruiting several people from the Signal Corps facility in Fort Monmouth, New Jersey, where TIROS was being built. They and the TIROS project were relocated to the Goddard Space Flight Center in Greenbelt, Maryland.[9] Finally, in May 1959 NASA created a Joint Meteorological Satellite Advisory Committee, its members drawn from NASA, the Defense Department, and the Weather Bureau, to coordinate meteorological satellite R&D efforts.[10]

TIROS was initially a three-satellite project, each satellite to have three instruments. The largest would be the television system for transmitting visible light images to the ground. The television system comprised two lenses: one wide angle imaging a square area about eight hundred miles on a side, and a narrow angle lens that photographed a much smaller area inside the same region. The second instrument was to be an improved version of an infrared sensor that had flown on Vanguard II. It could detect five different wavelength bands, allowing it to sense reflected solar radiation, water vapor absorption in the atmosphere, outgoing long-wave infrared radiation, and visible light. The third instrument, designed by scientists at the University of Wisconsin, would measure the earth's albedo to indicate heat balance. The first TIROS, however, flew with only the television camera.[11]

NASA placed the first TIROS into an inclined equatorial orbit from which it could photograph Earth between 50 degrees north and south latitude. The satellite began sending back images almost immediately, and over its seventy-six-day life it transmitted 22,952 useable photographs. TIROS showed an unreported typhoon near New Zealand, but its images revealed entirely unexpected structural features in clouds. TIROS photographs of oceanic storms clearly revealed their spiral-banded structure for the first time. They also showed that "mountain wave" cloud structures were much larger in scale than previously believed, extending from the Andes Mountains across the entire width of South America and displaying equally unexpected short- and long-wave structure.[12] Finally, TIROS images showed that unexpectedly rapid changes in the cloud patterns of vortexes occurred in the early phases of storm system formations.

At the third meeting of the Space Science Board's Committee on the Meteorological Aspects of Satellites in June, Harry Wexler, the Weather Bureau's chief of research, told the group that the TIROS images had triggered a review of convective cell research to try to explain why cloud structure was so much more variable than previously believed and why clouds assumed clear organization over many different scales.[13]

In addition to the scientific interest it generated, TIROS I was also enormously popular in the public arena and the more important political realm. TIROS provided results far more visible to average people (and politicians) than such other early space-age discoveries as the Van Allen radiation belts. Its photographs of Earth's cloud cover placed a phenomenon of everyday life into a new context. The photographs of oceanic clouds suggested to anyone interested in the weather that TIROS could photograph storms at sea before they reached land, significantly lengthening storm-warning times. The satellite received a four-column front-page article in the *New York Times*, which also reproduced two of the satellite's first photographs.[14] Walter Sullivan, the paper's science writer, pointed out that TIROS presaged an "era when such vehicles [would] produce a constant stream of information of great, immediate value in our daily lives."[15] People easily perceived the benefits of such "storm patrol" satellites. Other outlets followed the *Times*' lead, making TIROS and its instantly recognizable cloud photographs famous.

METEOROLOGICAL COLLABORATION IN THE COLD WAR

The success of TIROS I produced a flood of international interest. Many nations stood to gain from TIROS cloud imagery, particularly those with coastal regions that would benefit from longer storm warning times. That interest sparked a two-pronged process to incorporate meteorological satellites into the growing international meteorology infrastructure. In 1958, following the launch of the Soviet *Sputnik*, the United Nations established a Committee on the Peaceful Uses of Outer Space. After TIROS I, the General Assembly adopted Resolution 1721, which required the World Meteorological Organization to propose a program of study to "advance the state of the atmospheric sciences and technology in order to improve existing weather forecasting capabilities."[16] The WMO's response was a report calling for the construction of a World Weather Watch, essentially formalizing its expanding observational network, while asking an independent body, the International Council of Scientific Unions, to formulate a global atmospheric research program.[17] The WMO intended that the research program help integrate the new satellites into world forecasting systems.

The principal architects of the WMO's report were the Soviet academician V. A. Bugaev and the U.S. Weather Bureau's chief of research, Harry Wexler. Wexler had been personally involved in postwar meteorology's first great advance, the construction of workable numerical weather prediction models during the late 1940s and through the 1950s. He was, as historian Clark Miller has pointed out, a significant promoter of scientific internationalism via his International Geophysical Year efforts.[18] He was also a supporter of satellite-based meteorology, funding weather satellite research at the University of Wisconsin and establishing the Weather Bureau's satellite center. Until his untimely death in 1963, he served as a highly respected advocate of satellite meteorology.

At the same time Wexler was orchestrating the WMO's report, American political leaders were attempting to interest the Soviet Union in cooperative space research. At the request of President John F. Kennedy's science adviser, MIT engineer Jerome Wiesner, NASA developed a set of proposals for cooperation that ranged from the joint use of tracking and data facilities to a unified effort to land men on the moon.[19] Specifically included among the proposals was cooperation in meteorological satellites, and this was the plan that the two nations agreed upon. So-called World Data Centers, one in Washington and one in Moscow, linked by dedicated telecommunications circuits, would receive and archive data from the two nations' meteorological satellites. Meetings in 1962 and 1963 worked out the technological details of the bilateral system, around which a series of World Meteorological Organization reports between 1963 and 1966 defined the structure of a world network.

In deciding to participate in a cooperative program with the Soviet Union and to support the World Meteorological Organization's effort to produce a global atmospheric observation system, the American government had essentially decided to make available the benefits of its most advanced technology (although not, of course, the technology itself) to the world's nations. That decision was also reflected in the equipment aboard the satellites. The initial TIROS had used a vidicon-based photographic system whose data was recorded on tape aboard the satellite. It was transmitted to the ground only when commanded to by one of two ground stations, located in Alaska and Virginia. The relatively high data rate required for a ground station to capture one orbit's imagery while the satellite was within its line of sight resulted in an expensive ground station, and one that utilized technologies that the United States did not then wish to share. The ground station also required a large antenna that rendered it particularly immobile.

These factors pushed the U.S. Navy, which wanted its ships able to receive imagery relevant to their operating areas, to develop a different system that could utilize a small, inexpensive, "low-tech" ground station compatible with shipboard use. Instead of storing all imagery on tape for transmission upon command, this system transmitted each image as it was taken. Eliminating the need for two-way communication between satellite and ground had a price: Images would be lost if no ground station happened to be in range. The image resolution was also lower than that of the earlier TIROS system, to reduce the complexity of the receiver. Negotiations between the navy, NASA, and the Weather Bureau got this system released for use in the civilian weather satellite program under the name Automatic Picture Transmission. The first launch of an APT-equipped TIROS came in 1966.

Weather Bureau and NASA leaders had intended the APT system to serve both the navy and individual users worldwide, and they published detailed "how-to" articles in trade journals to encourage hobbyists to build their own receivers. The parts cost for the ground stations was about three thousand dollars, and the parts were relatively common. While not inexpensive relative to a household budget, it was easily affordable to most national governments. APT technology allowed the United States to share the benefits of its space investment with most nations without having to give valuable technologies away. It also aided NASA's efforts to promote space applications and fostered the government's larger aim of demonstrating American commitment to internationalism.

But while APT was a visible American commitment to internationalism, it was not really part of meteorologists' drive to create global weather prediction. Cloud imagery's principal advantage to forecasters was advanced warning via early identification of oceanic storms. The global forecasting that meteorologists sought, however, was the ability to predict the occurrence and movement of all types of weather phenomena. At John von Neumann's famous Meteorology Project at the Institute for Advanced Study in Princeton, Jule Charney and others had created computer models of the atmosphere to predict the weather.[20] These models required wind and temperature measurements from the real atmosphere, which were provided during the 1950s by networks of weather balloon stations in North America and Europe. These did not exist in ocean regions, the tropics, or in the Southern Hemisphere, limiting the early numerical experiments to regions in the Northern Hemisphere.

The leaders of the American numerical modeling effort had global ambitions. They and their European colleagues drove the effort to expand the observation network so that it could eventually feed global models. This was behind their effort to create the World Weather Watch system, and it also led to their proposing a global atmospheric research program. A radiosonde network could not cover the oceans or poles, both of which were important to forecasting. Charney and his colleagues believed that satellites more advanced than TIROS might be able to provide them with measurements from these "data sparse" regions. These satellites needed to be developed and the quality of the data they produced checked against other sources of data before they could be accepted into a forecasting system, and this need had been behind the World Meteorological Organization's call for a meteorological research program parallel with the construction of the World Weather Watch.

The organization tasked with creating the research program, the International Council of Scientific Unions, relied on the "scientific union" of its member states to make proposals that would be the basis of negotiation and agreement. In the United States, the relevant scientific union was the National Academy of Sciences, which asked Jule Charney to organize the proposal. Titled *The Feasibility of a Global Observation and Analysis Experiment*, Charney's 1966 proposal became known as the "Blue Book" for the color of its cover—in the words of one scientist, "the gospel according to St. Jule."[21]

Charney's proposal divided the research program into two problem areas: exploitation of space and data processing technologies to provide global observations, and improvement of scientific understanding of turbulent transport of matter and energy in the atmosphere. It also specified the need for several subprograms. Tropical circulation was poorly understood. As the tropics were where most of the sun's radiation reached Earth, study of tropical energy flows was vital. Energy exchange between the atmosphere and land and ocean surfaces was also a poorly understood process, but essential to accurate long-range weather prediction; an observational program to determine the dynamics of these energy flows was vital. Finally, design studies of a global observing system that could meet the scientific needs of the research program had to be carried out. Charney's plan was slightly modified during a July 1967 meeting in Stockholm, and it became known as GARP.

GARP, as it was carried out during the 1970s, consisted of a series of field experiments, beginning with the tropical expedition Charney had proposed and ending with a global experiment designed to produce detailed, validated data sets that would be, in essence, sets of snapshots of the global atmosphere that could be used to improve the prediction models. The first field experiment, known originally as TROMEX (Tropical Meteorology Experiment), was carried out during the summer of 1974 as the GARP Atlantic Tropical Experiment, or GATE. GATE had a fleet of forty ships, thirteen aircraft, and several (American) satellites devoted to its three observational periods.[22] The ships were positioned across the equatorial Atlantic in a set of nested arrays to observe the atmosphere at different scales: large-scale, synoptic motion over a thousand or more kilometers; medium-scale motion on the order of hundreds of kilometers; and small-scale phenomena, a hundred or fewer kilometers. An American geosynchronous imaging satellite parked over the ship array provided half-hourly photographs of the region, permitting almost-real-time adjustment of aircraft flight plans to accommodate the ever-changing weather.

Originally, GATE was to have been conducted simultaneously with a set of tests on functional prototypes of the sensors NASA and the U.S. National Oceanic and Atmospheric Administration (NOAA) intended for TIROS-N, the next generation of operational meteorological satellites planned for the late 1970s. One sensor, known as the TIROS Operational Vertical Sounder (TOVS), detected infrared and microwave radiances of atmospheric gases in order to derive temperature profiles of the atmosphere. These profiles were necessary to initiate global forecast models. Another, the Tropical Wind Experiment and Reference Level System, was essentially a tracking and data relay system used in conjunction with balloons and floating ocean buoys. Its purpose was to provide wind and surface pressure measurements in the tropics. These were to be launched in June 1974 on the sixth of a series of NASA research satellites named Nimbus. Delays to one of the primary sensors pushed the launch into 1975, however, and the tests were also delayed.

These "Data Systems Tests" were designed as end-to-end tests of the new instruments, and the data processing system that received their raw data, transformed it into meteorological variables, and generated analyses. They had been scheduled simultaneously with GATE because Nimbus 6 carried two experiments that complemented some of GATE's five hundred subexperiments. A constant-level, satellite-tracked balloon system called the Tropical Wind, Energetics, and Reference Level experiment would provide wind and temperature measurements in the lower stratosphere as well as demonstrating the capabilities of the tracking system. These measurements could be compared to those produced from the satellite's vertical temperature profile instrument, which estimated temperature from the radiance of carbon dioxide molecules in the atmosphere. The other experiment complementing GATE was known as the Earth Radiation Budget, a set of sensors designed to measure the amount of energy radiated by the sun, reflected off Earth (mostly by clouds) and radiated by Earth. GATE contained a radiation subprogram designed to examine the radiative properties of the ocean surface and of clouds, and this instrument was intended to support that program as well as generate regional maps of Earth's radiance.

The data requirements of GATE were similar to those of the global experiment that GARP was working toward, making it a useful test scenario. Finally, because a major GATE goal was analysis of the interaction between different scales of atmospheric motion, global to local and vice versa, it still required a global data set. But there were other polar orbiting satellites available to provide global temperature measurements with older, lower-resolution experimental instruments than those carried by Nimbus 6.

Hence, delay to Nimbus 6 meant that some of GATE's subprograms were impaired, but it did not significantly affect the primary experiment goal of examining tropical energy exchange.

When the Data Systems Tests were carried out a year after GATE, the results were somewhat disturbing. Portions of the tests went very well. A satellite-tracked balloon experiment, for example, carried out launches of 393 instrumented, constant-level balloons to 150 mb altitude in the tropical Southern Hemisphere between October 1975 and January 1976. The tracking system performed essentially as expected, giving a location accurate to within about 3 km, permitting derivation of wind velocities from the balloons' motion. It also successfully tracked and received data from drifting buoys, an important part of the future global observation system.[23]

But the Data Systems Tests also demonstrated that future global observation system needed to have much greater quality control in the data processing path. When the Goddard Institute for Space Studies and the National Meteorological Center tried to use the satellite data in their experimental global models during 1976 and 1977, they found that it had damaging errors. One source of error came from the use of wind data derived from the motion of clouds, which could be tracked through successive images from geosynchronous satellites. The data sets were supposed to contain wind vectors at two altitudes, but operators had difficulty in consistently assigning clouds to altitudes.[24]

Similarly, the temperature profiles derived by the National Environmental Satellite Service from the temperature sounder on Nimbus 6 contained substantial errors. These errors derived from an inability to eliminate the effects of clouds on the temperature measurements under certain conditions. The sensor could not see through clouds, and clouds had the effect of producing an artificially cold temperature reading. William L. Smith of NOAA had developed an automated cloud removal method that worked most of the time, but it failed often enough to produce significant forecast errors. Writing for the record in November 1977 about the temperature sounding challenge, Thomas O'Neill, commenting for the U.S. Committee for GARP, concluded that the data processing methods used during the final global experiment should not be the same as those used during the Data Systems Tests. A special effort to check the quality of the sounding retrievals during the experiment was going to be necessary.[25]

During the early 1970s, Vern Suomi of the University of Wisconsin had invented a method of partially automating the cloud-tracking process that allowed derivation of wind velocities from geosynchronous satellite imagery. Initially called WINDCO and later McIdas, the system utilized a minicomputer as a workstation, a remote mainframe for processing, and a television-quality monitor that was linked via a "lookup table" to the much higher-resolution image data. Operators could use the high-resolution data without actually having to display it—video displays with sufficient resolution to display the 120-megabyte satellite images did not exist.

One of the principal difficulties of tracking clouds in successive images had been that the satellite's own motion was imperfect, and it had to be subtracted from the overall cloud motion. WINDCO did this using two corrections, one based on the satellite's known motion from the satellite tracking system, and one based on registration of the images to an obvious landmark on Earth. The WINDCO operator chose a landmark visible in successive images with a light pen, and the computer made the necessary corrections to its model of the satellite's motion. The operator could then select clouds for tracking and produce the wind set. Suomi liked to call this method of combining human intelligence with automation "man-interactive computing," and this was the genesis of the system's final name, the "Man-Computer Interactive Data Access System." McIdas, demonstrated for NASA and NOAA officials in April 1972, became the means of production of the cloud-tracked wind sets for GATE and for the Data Systems Tests in 1974 and 1975.[26] The SSEC was assigned responsibility for receiving and archiving the satellite imagery during these experiments and producing four sets of wind data each day.

In January 1978 the First GARP Global Experiment (FGGE) Advisory Committee met in Madison to work out how NASA, NOAA, and the SSEC would deal with the problems. At this meeting, NOAA's William L. Smith argued that the satellite sensors already provided most of the data necessary to produce better outcomes.[27] Assignment of altitudes to the clouds tracked to produce winds could be done more accurately if the McIdas operators had access to the cloud-top temperature data provided by the "window" channel since cloud temperature was directly related to altitude. Similarly, McIdas operators could crosscheck temperature soundings with the National Meteorological Center's analysis charts, and with cloud imagery to evaluate them if these were made available to the operator. In this way, trained meteorologists' subjective judgment could be used to check the performance of the automated cloud-clearance process.

While having to use humans to inspect the results of the automated sounding results slowed the process somewhat, only a small fraction of the twenty thousand soundings generated by the TIROS-N's temperature instrument each day would require human intervention. A set of automated filters that compared soundings to nearby radiosondes already threw out obviously bad measurements. Another set of filters that compared the soundings from the infrared instrument to those from the microwave instrument was under development at the National Meteorological Center. Because the two instruments had significantly different resolution, instead of automatically discarding soundings that differed, these filters would flag them for inspection. Under certain meteorological conditions, one would expect the large-area microwave sounding to differ from the higher-resolution infrared result without either of them being incorrect. Meteorologists could identify these and validate or reject the sounding. This was another area in which McIdas's ability to display multiple data sources graphically would help speed the process.

As a result of this meeting, the committee recommended that the SSEC develop and implement McIdas software to enable its use for these quality control measures. In a "special effort," the FGGE data sets would be checked by meteorologists using McIdas; later in the year, the SSEC proposed for and won this task. It also built a McIdas system for NOAA's National Environmental Satellite Service for operational use. As a result of these recommendations, the GARP project office restructured the data flow path for the FGGE: The SSEC's McIdas terminal would be linked to an Amhdahl mainframe at the Goddard Institute for Space Studies in New York for the data processing and to the National Meteorological Center so that forecast analyses could be imported directly into McIdas.[28]

GARP's planners were also disturbed by the National Meteorological Center's assessment that the satellite temperature profiles did not significantly improve forecast ability in the Northern Hemisphere.[29] This determination was independent of the problem of erroneous soundings; even when they were weeded out and discarded, the center's forecast model produced essentially the same results as it did when given only the conventional radiosonde data. In the Southern Hemisphere, where almost no conventional data existed, the satellite data improved forecasts significantly, bringing Southern Hemisphere forecasts to nearly the same level as those in the Northern Hemisphere. This ratified the results of a series of Observing Systems Simulation Experiments carried out between 1970 and 1973, which had suggested that 1970s generation of satellite instruments would only result in skillful four- to five-day forecasts.[30] But the GARP planners, most of whom represented Northern Hemisphere nations, had expected that satellite data would extend Northern Hemisphere forecasts beyond the four to five days possible with conventional data. Instead, the Data System Tests had suggested that the satellite soundings produced no benefit to their nations at all. This undermined the program significantly, both internationally and within the United States. GARP was an expensive program, and the United States was paying half its costs.

The First GARP Global Experiment (renamed the Global Weather Experiment when it became clear there would be no second global experiment) was carried out in 1979, three years after its original schedule. Its principal goal was the collection of two sixty-day sets of validated, quality-controlled global data for use in improving global atmospheric models. By this time, the global observing system consisted of five geosynchronous imaging satellites (three American, one Japanese, and one European), two polar orbiting satellites (both American), the conventional World Weather Watch ground/radiosonde network, and a set of temporary "special observing systems": a tropical wind-finding system based on satellite-tracked balloons, a drifting buoy system in the Southern Hemisphere also using satellite data relay and tracking, and expanded use of data provided by commercial aircraft and shipping.[31] The data sets, processed by NOAA and checked by the University of Wisconsin, became available internationally through the world data centers.

The research conducted using the FGGE/GWE data, unfortunately, validated the Data System Test results.[32] NASA, NOAA, and their university partners had fixed the quality control problem through the employment of the "man-interactive" computing system, but they could not overcome the limitations of the satellite instruments to accurately define the atmosphere's temperature structure. The best forecasts the global observation system could produce were limited to four to five days in the middle latitudes. This was not a great improvement for the developed nations of the Northern Hemisphere, which had gained forecasts of this length from the conventional observational network combined with the hemispheric models developed during the 1960s. In the tropics, the benefits were fewer, with forecast skill remaining, at best, two days. This was due to the lack of direct wind measurements; the models relied on an approximation to generate wind velocities from temperature measurements, but it did not work in the tropics. Better tropical forecasts meant developing and deploying a permanent wind measurement system. The developed nations paying for GARP did not see sufficient benefit to themselves in maintaining such a wind system after the global experiment.

LEGACIES OF GARP

After the Global Weather Experiment, enthusiasm for GARP, and for large-scale international scientific endeavors like it, dissipated rapidly in the United States. The last GARP field experiment, the Alpine Experiment (ALPEX), took place in Europe in 1982. Almost simultaneously, at the UNISPACE meeting in Vienna, NASA administrator James Beggs tried to initiate a global climate observing system under the rubric of "Global Habitability." A scientific panel led by Richard Goody had framed the scientific questions in need of research earlier in the year, and this meeting had been the concept's first public outing.[33] But it was soundly rebuffed, for several reasons. At the same time NASA was proposing this new initiative, the Reagan administration's Commerce Department was trying to commercialize the Landsat and weather satellite systems.[34] This would force national governments to start paying for data that had been free, a reversal of the Kennedy administration's long-standing policy. While this effort failed eventually, and the weather satellite data remained free, it damaged American credibility in the short term.

Further, GARP had itself started as a bilateral arrangement with the Soviet Union, but that was not possible in the political climate of 1982. The Reagan administration had quite deliberately taken a hostile, aggressive posture toward the Soviet Union, jump-starting the Cold War after the decade of détente. Because the Soviet Union had great sway in the Third World, its lack of support helped ensure Third World nations' hostility to the program. Finally, NASA's leaders had not done the necessary work to bring other potential partners on board prior to making the announcement.[35] GARP had spawned the World Climate Research Program in 1980, but NASA had not worked with it. NASA had also not sought the support of its traditional European partners. For all these reasons, the NASA initiative was seen as unilateral effort, not an international one, and the Global Habitability Project received a very cold reception.

It did not die, of course. Submerged from view and nurtured by NASA's bureaucracy, the agency eventually did all the things it should have done in the first place and brought the idea back to life as the Mission to Planet Earth in 1986.[36] Similarly, the scientific internationalism that had motivated GARP survived. But the mechanisms by which it operated changed considerably. GARP had been a truly international undertaking, administered via the United Nations and the International Council of Scientific Unions; for its climate research program, NASA pursued a multinational approach. It sought specific partners for individual missions, and it kept its few field campaigns during the decade restricted in size, scope, and participation. While this design was ostensibly to prevent the project's capture by international institutions, it also helped insulate the project from domestic adversaries of international science.[37] Further, NASA's leaders believed that GARP had not been very productive scientifically. The in situ measurement, satellite and modeling efforts had not been tied closely enough together, and the scientific effort was diffused across too many disciplines, organizations, and researchers to make effective progress.[38]

Yet GARP did succeed at stimulating global modeling. During the first half of the 1980s, model researchers made significant improvements in meteorological models using the GWE data set.[39] New modeling centers also emerged, with at least one surpassing the U.S. National Meteorological Center in the skill of its operational forecasts, the United Kingdom's Meteorology Office (Met Office). NASA's two late-1980s stratospheric ozone expeditions to the Antarctic and the Arctic relied on Met Office polar analyses despite being almost entirely American affairs, while the lack of conventional data sources at the ends of the world made these scientific expeditions dependent on the American satellite infrastructure.[40] But weather modeling also spread outside the developed world. China and India both developed model-based forecast centers, and the Chinese government bought a copy of the University of Wisconsin's data-processing system as part of its equipment. India eventually developed its own geosynchronous weather satellite, although it did not share the data with the World Weather Watch system. It considered the data too important to its enemies to give away.

The United States also gained from its fostering of global meteorology. American domestic forecasts improved due to the Weather Bureau's access to overseas data. A study done using the GWE data set for the Air Force Geophysics Laboratory during the mid-1980s, for example, demonstrated that loss of conventional data from the Soviet Union would impair North American forecasts measurably within four days.[41] Because the weather is a global phenomenon, it was in the interests of the American government to leave the construction and functioning of the World Weather Watch system independent of the changing tenor of Cold War politics. But even without foreign conventional data, the satellite-based global forecasting the United States had built for itself during GARP could maintain global three- to four-day forecasts of useful skill indefinitely—something it could not have done prior to the space age.

The legacies of GARP, then, are complex. It fostered data exchange and numerical modeling efforts throughout the world. It helped demonstrate the value of satellite technology for forecasting, and its data sets helped identify what future progress in forecasting would require. It also helped demonstrate the continued value of keeping humans involved "in the loop." It fell far short of its specified goal of monthly forecasts, while it helped demonstrate these were actually impossible to achieve anyway. It strengthened the American atmospheric research community's overseas ties, even if GARP itself was not replicated. It provided foundations for cooperation in climate research, although these took years to mature. And it gave the United States a global forecasting ability that served the needs of its world-spanning military.

NOTES

1. On the early failures of the space age, see Walter A. McDougall, *The Heavens and the Earth: A Political History of the Space Age* (New York: Basic Books, 1985), 141–76.

2. World Meteorological Organization, *The WMO Achievement: 40 Years in the Service of International Meteorology and Hydrology*, Report 729 (Geneva: WMO, 1990).

3. Clark A. Miller, "Scientific Internationalism in American Foreign Policy: The Case of Meteorology, 1947–1958," in *Changing the Atmosphere: Expert Knowledge and Environmental Governance*, ed. Clark A. Miller and Paul N. Edwards (Cambridge, MA: MIT Press, 2001), 167–217.

4. Paul N. Edwards, *The Closed World: Computers and the Politics of Discourse in Cold War America* (Cambridge, MA: MIT Press, 1996), 8–12.

5. Charles C. Bates and John F. Fuller, *America's Weather Warriors, 1818–1985* (College Station: Texas A&M University Press, 1986), chap. 7.

6. Kristine C. Harper, "Boundaries of Research: Civilian Leadership, Military Funding, and the International Network Surrounding the Development of Numerical Weather Prediction in the United States" (PhD thesis, Oregon State University, 2003); see also Frederik Nebeker, *Calculating the Weather: Meteorology in the 20th Century* (San Diego, CA: Academic Press, 1995); William Aspray, *John von Neumann and the Origins of Modern Computing* (Cambridge, MA: MIT Press, 1990).

7. R. Cargill Hall, *A History of the Military Polar Orbiting Meteorological Satellite Program* (Washington, DC: National Reconnaissance Office, 2001), 2.

8. Richard Leroy Chapman, "A Case Study of the US Weather Satellite Program: The Interaction of Science and Politics" (PhD thesis, Syracuse University, 1967), 33.

9. Chapman, "A Case Study," 70–80.

10. John Allen to T. Keith Glennan, 30 November 1960 (Wexler papers, box 31, file 10); U.S. Senate, *NASA Authorization for Fiscal Year 1961*, part 2, *Hearings*, 86th Cong., 2nd Sess., 30 June 1960, 646.

11. National Aeronautics and Space Administration, *Significant Achievements in Satellite Meteorology, 1958 to 1964* (Washington, DC, NASA SP-96, 1966), 12, 56.

12. Space Science Summer Study, Working Group on Meteorology, 29 June 1962 (Berkner papers, box 68, "SSB-SS-Meteorological Rockets and Satellites"). See also S. Fritz and H. Wexler, "Cloud Pictures from Satellite TIROS I," *Monthly Weather Review* 3, no. 88 (March 1960): 79–87.

13. Minutes, Committee on Meteorological Aspects of Satellites, 24–25 June 1960 (SSB collection, file "Committees, Meteorological Aspects of Satellites, Third Meeting").

14. "U.S. Orbits Weather Satellite, It Televises Earth and Storms: New Era In Meteorology Seen," *New York Times*, 2 April 1960, 1.

15. "U.S. Orbits Weather Satellite," 4.

16. International Council of Scientific Unions, Committee on Atmospheric Sciences, "Report of the Study Conference held at Stockholm, 28 June–11 July 1967, on the Global Atmospheric Research Program," 1967, ii.

17. World Meteorological Organization, *Final Report on the Advancement of Atmospheric Sciences and their Application in the Light of Developments in Outer Space* (Geneva: WMO, 1962).

18. Miller, "Scientific Internationalism in American Foreign Policy," 167–218.

19. Special Assistant for Science and Technology (Wiesner) to President Kennedy, 20 Feb. 1961 (Kennedy Library, National Security Files, Departments and Agencies Series, Space Activities, General 1/61–3/61, Box 307), reproduced in *Foreign Relations of the United States*, vol. 25, *1961–1963*, ed. Paul Claussen, Evan M. Duncan, and Jeffrey A. Soukup (Washington, DC: Government Printing Office, 2001), doc. no. 386; "Draft Proposals for US-USSR Space Cooperation," 4 April 1961, reproduced in *Exploring the Unknown: Selected Documents in the History of the U.S. Civil Space Program*, vol. 2, *External Relationships*, ed. John M. Logsdon, Dwayne A. Day, and Roger D. Launius (Washington, DC: NASA, 1996), 143–46; "Topical Summary of Bilateral Discussions with Soviet Union," 27–30 March 1962, reproduced in *Exploring the Unknown*, 153–62.

20. The most detailed study of Von Neumann's meteorological project is Harper, "Boundaries of Research." See also Aspray, *John von Neumann*, 143–46.

21. National Academy of Sciences–National Research Council, *Feasibility of a Global Observation and Analysis Experiment*, Publication no. 1290 (Washington, DC: Author, 1966).

22. Joachim P. Kuettner, "GATE Final International Scientific Plans: General Description and Central Program of GATE," *Bulletin of the American Meteorological Society* (hereafter *BAMS*) 55, no. 7 (July 1974): 711–19; D. E. Parker, "The Synoptic-Scale Subprogram," *BAMS* 55, no. 7 (July 1974): 720–23; D. R. Rodenhuis, "The Convection Subprogram," *BAMS* 55, no. 7 (July 1974): 724–731; H. Hoeber, "The Boundary-Layer Subprogram for GATE," *BAMS* 55, no. 7 (July 1974): 731–34; H. Kraus, "The Radiation Subprogram of GATE," *BAMS* 55, no. 7 (July 1974): 734–37; S. G. H. Philander, "The Oceanographic Subprogram of GATE," *BAMS* 55, no. 7 (July 1974): 738–44; Joachim P. Kuettner and David E. Parker, "GATE: Report on the Field Phase," *BAMS* 55, no. 7 (July 1974): 11–28.

23. On TWERLE, see TWERLE Team, "The TWERL Experiment," *BAMS* 58, no. 9 (September 1977): 936–48. The article is, unusually for this journal, attributed to the team instead of to the individual authors because the authors wished to dedicate it to two University of Wisconsin graduate students, Charles Blair and John Kruse, who perished in an aircraft accident during preparation for the experiment. The TWERLE science team was Paul Julian, Vincent Lally, William Kellogg, Vern Suomi, and Charles Cote.

24. Thomas H. R. O'Neill to the Record, 28 November 1977 (file "USC-GARP FGGE Advisory Panel, FGGE Quality Control," box no. 84–009–2, National Academies Archive).

25. O'Neill to the Record.

26. Space Science and Engineering Center, *Space Capsule* 2, no. 3 (April 1972).

27. "Report on Informal FGGE Quality Control Meeting," typescript, 10–12 January 1978 (Box 84–009–2, file "USC-GARP FGGE Advisory Panel, FGGE Quality Control," National Academies Archives).

28. James R. Greaves, et al., "A 'Special Effort' to Provide Improved Sounding and Cloud-Motion Wind Data for FGGE," *BAMS* 60, no. 2 (February 1979): 124–27.

29. A. Desmarais et al., "The NMC Report on the Data Systems Test," Camp Springs, MD, March 1978; Thomas H. R. O'Neill to Verner E. Suomi, 12 December 1977 (Box 84–009–2, file "USC-GARP FGGE Advisory Panel, FGGE Quality Control," National Academies Archives). See also M. Steven Tracton and Ronald D. McPherson, "On the Impact of Radiometric Sounding Data upon Operational Numerical Weather Prediction at NMC," *BAMS* 58, no. 11 (Nov. 1977): 1201–1209; Tracton, A. J. Desmarais, R. J. Van Haaren and McPherson, "The Impact of Satellite Soundings on the National Meteorological Center's Analysis and Forecast System—The Data Systems Test Results," *BAMS* 108, no. 5 (May 1980): 543–86.

30. Milton Halem and Robert Jastrow, "Analysis of GARP Data Requirements," *Journal of the Atmospheric Sciences* 27, no. 1 (January 1970): 177; Jastrow and Halem, "Simulation Studies Related to GARP," *BAMS* 51, no. 6 (June 1970): 490, 498–501.

31. Rex Fleming et al., "The Global Weather Experiment: 1. The Observational Phase through the First Special Observing Period," *BAMS* 60, no. 6 (June 1979): 649–59.

32. National Research Council, *Proceedings of the First National Workshop on the Global Weather Experiment* (3 vols.; Washington, DC: National Academies Press, 1985).

33. Jet Propulsion Laboratory, "Global Change: Impacts on Habitability. A Report by the Executive Committee of a Workshop held at Woods Hole, MA 21–26 June 1982," JPL document D-95, 1982. Also see National Research Council, *Earth Observations from Space: History, Promise, Reality* (Washington, DC: National Academy Press, 1995), chap. 2.

34. M. Mitchell Waldrop, "Imaging the Earth (I): The Politics of Landsat," *Science* (26 March 1982): 1600–1603; Waldrop, "Imaging the Earth (II): The Politics of Landsat," *Science* (2 April 1982): 40–41; Waldrop, "What Price Privatizing Landsat?" *Science* (11 February 1983): 752–54. For a critical review of this history, see National Research Council, *Earth Observations from Space*), 109–15.

35. See Shelby Tilford interview with Erik Conway, 11 February 2004. For a slightly different perspective, see Office of Technology Assessment, *Unispace 1982: A Context for International Cooperation and Competition* (Washington, DC: Government Printing Office, 1983).

36. For an outline of the history of MTPE, see National Research Council, *Earth Observations from Space*, chap. 2.

37. On the management aspect of Global Habitability, see Kenneth Pederson to Associate Administrator for Space Science and Applications (Burt Edelson), "Global Habitability as a U.S. Initiative at UNISPACE 82," 1 July 1982 (NASA History Office, Earth Observing System Collection, box 6).

38. See Tilford interview.

39. William L. Smith, "Atmospheric Soundings from Satellites—False Expectation or the Key to Improved Weather Prediction," *Quarterly Journal of the Royal Meteorological Society* 117 (1991): 267–97.

40. See A. F. Tuck, R. T. Watson, E. P. Condon, J. J. Margitan, and O. B. Toon, "The Planning and Execution of ER-2 and DC-8 Aircraft Flights Over Antarctica, August and September 1987," *Journal of Geophysical Research* 94, no. D9 (30 August 1989): 11181–222.

41. J. F. Louis et al., "Baseline Observing System Experiments Using the AFGL Global Data Assimilation System," Report no. AFGL-TR-87-0327 (Cambridge, MA, 18 November 1987), 20.

Index

About the Editors and Contributors

Ann M. Becker received her doctorate from Stony Brook University in 2005, and also holds an MA in history and an MA in teaching social studies from Stony Brook. Her primary academic focus is college-level teaching, both history and writing. Dr. Becker's primary research interests include the military impact of smallpox during the Revolutionary War, as well as Revolutionary pensions, and she serves as the associate editor for the *Long Island Historical Journal*.

Erik M. Conway is a historian of science and technology at the Jet Propulsion Laboratory, California Institute of Technology, Pasadena, California. He is the author of *High-Speed Dreams: NASA and the Technopolitics of Supersonic Aviation* (2005), and *Blind Landings: Low Visibility Operations in American Aviation* (2006), both from Johns Hopkins University Press. He has recently completed a history of atmospheric science in NASA.

Robert G. Ferguson is researching and writing a book on the history of NASA's aeronautics research. He has been a Ramsey Fellow at the Smithsonian's National Air and Space Museum and was on the faculty at the Hong Kong University of Science and Technology, where he was awarded the Michael G. Gale Medal for distinguished teaching. The primary focus of his research to date has been the American airframe industry in World War II, with a special interest in the relationships among organizational innovations, manufacturing, and technical development.

James R. Fleming teaches history of science and technology at Colby College. His books include *Meteorology in America, 1800–1870* (Johns Hopkins, 1990),

Historical Perspectives on Climate Change (Oxford, 1998), and *The Callendar Effect* (American Meteorological Society, 2006). He held the Charles A. Lindbergh Chair in Aerospace History at the National Air and Space Museum in 2005–2006 and was awarded the Roger Revelle Fellowship in Global Stewardship from the American Association for the Advancement of Science in 2006–2007. He is a public policy scholar with the Woodrow Wilson International Center for Scholars, where he is working on a history of weather and climate control.

Barton C. Hacker works with the military history collections of the Smithsonian's National Museum of American History, but museum curator is a relatively recent redirection of his career. Before coming to Washington in 1998, most of his professional activity centered on research and writing (and sometimes teaching) about the historical interactions between military institutions and other social institutions as often mediated by military technological change. His major publications have dealt with the history of military technology, women's military history, and the comparative world history of military institutions.

John M. Morra is retired from teaching history in suburban Cincinnati. He taught at the University of Cincinnati and Eastern Kentucky University. His major field of interest was the history of ideas, before he encountered the career of Dr. Albert Sabin. At present, he is working on a biography of Sabin.

Walter Earl Pittman is president of Fort Stanton, Inc., the historic preservation group restoring the old cavalry fort. A retired professor of history, he holds three degrees in history and three in chemistry, geology, and physics. He has written extensively on military and environmental history and is writing on the Civil War. A veteran of U.S. Special Forces, he also served as an intelligence officer in the U.S. Navy and U.S. Air Force and wrote the history of Special Operations during the Cold War.

Mark Soderstrom is an assistant professor at SUNY–Empire State College. He has published articles in *Journal of the Gilded Age and Progressive Era* and *XCP*, and is revising a manuscript on racial science and segregation in Big Ten universities tentatively titled *Weeds in Linnaeus's Garden.* His research interests include oral history, labor history, and the history of science and race.

Keir B. Sterling was a college history professor, administrator, and consultant for over twenty years before serving as command historian and instructor at the U.S. Army Ordnance Center and School, Aberdeen Proving Ground, Maryland, from 1983 to 1998. Since then, he has been historian for the U.S. Army Combined Arms Support Command and is on the teaching staff of the Army Logistics Management College at Fort Lee, Virginia. His research and publications have focused on historical developments in vertebrate zoology and conservation, the U.S. government's role in natural science, and military logistics. He was senior editor (with R. Harmond, G. Cevasco, and L. Hammond) of *Biographical Dictionary of American and Canadian Naturalists and Environmentalists* (Greenwood, 1997).

Margaret Vining is a curator in military and diplomatic history at the National Museum of American History. She is in charge of collections acquisition and management, preservation, research, writing, and museum exhibitions of United States military and diplomatic material culture. She has in the past fifteen years served as principal curator or cocurator of numerous major exhibitions as well as important small showcases. Among her specialties and the subjects of many of her lectures and publications are women's military history, the social history of military technology, World War I, and military uniforms.